The Genesis of
Quantum Theory
(1899-1913)

The MIT Press
Cambridge, Massachusetts,
and London, England

**The Genesis of
Quantum Theory
(1899-1913)**

Armin Hermann

Translated by
Claude W. Nash

Originally published by Physik
Verlag, Mosbach/Baden, under the
title "Frühgeschichte der Quantentheorie (1899-1913)"

English translation
Copyright © 1971 by
The Massachusetts Institute of
Technology

This book was designed by The
MIT Press Design Department.
It was set in IBM Composer Press
Roman
by The Science Press, Inc.,
printed on Finch Textbook Offset
by Halliday Lithograph Corp.,
and bound in Joanna Oxford book-
cloth
by The Colonial Press Inc.
in the United States of America.

ISBN 0 262 08047 8 (hardcover)

Library of Congress catalog card
number: 73–151106

118757

Preface

Without doubt the most significant advance in physics in our century is the genesis of the quantum theory. Its influence is by no means confined to physics and chemistry but involves fundamental questions of human knowledge itself. For this reason the historical background of the theory, quite aside from its physical and philosophical aspects, has been the object of early investigations. F. Hund, M. Jammer, M. J. Klein, and L. Rosenfeld have all published papers on this development. In recent years a large-scale project, "Sources for History of Quantum Physics" (see page 162) has secured a gigantic body of source material, above all letters of the scientists involved in the early stages of quantum physics, and has collected it in the form of microfilms.

The present work attempts for the first time to make extensive use of unpublished sources dating back to the first fifteen years of this development, from 1899 to 1913. In addition to the letters and manuscripts collected by the "Sources," a great deal of previously unknown material has been included. In this effort the manuscript collections of libraries and archives were widely consulted, as were particularly the scientific legacies of Johannes Stark and Arthur Erich Haas as well as the most essential portion of Arnold Sommerfeld's papers. It was the intention to conjure up as vivid as possible an image of the genesis of the quantum theory by drawing on pertinent publications and letters of the years 1899 to 1913, at the same time making cautious use of the inevitably somewhat distorted accounts from memory.

The present English edition contains all the references considered necessary for a scientific publication. On the other hand, the extensive literature survey that was originally published in the German version, a thesis for habilitation, has here been omitted since it is largely superfluous for the present purpose.

Thanks are expressed to the able translator, Mr. Claude W. Nash. It is hoped that this book will bring pleasure to the reader and gather new friends for the history of physics.

Armin Hermann
Stuttgart, September 1970

Preface
v

Introduction   The Continuity
1   Principle

1
Max Planck   New Physical Constants
5   and the Radiation Law
                           (1899-1906)

                           1. The Radiation Equilibrium
                           2. Derivation of Wien's Equation
                           3. The New Physical Constants
                           4. Planck's Radiation Equation
                           5. The Boltzmann Method
                           6. The Meaning of the Natural
                              Constants
                           7. "One of the Greatest Discov-
                              eries since Newton"

2
H. A. Lorentz   Discussion of the Radiation
29   Problem
                           (1903-1910)

                           1. The Reception of Planck's
                              Radiation Theory
                           2. Struggles with the Problem
                           3. The Rome Lecture of 1908
                           4. Adverse Criticism
                           5. The Dialogue between Planck
                              and Lorentz
                           6. Lorentz—A Leader in Science?

3
Albert Einstein   Light Quanta and New
50   Quantum Phenomena
                           (1905-1910)

                           1. The "Technical Expert, Third-
                              Class"
                           2. The Light-Quantum Hypothesis
                           3. Further Evidence Supporting
                              the Hypothesis

4. A Unified Theory of Light
Quanta and Electrons
5. The "Groups of Phenomena
Related to the Transformation
of Light"
6. The Quantum Theory of
Specific Heat
7. The Salzburg Congress of 1909

4

Johannes Stark

72

The Search for New Quantum
Phenomena

(1907-1910)

1. The Motives
2. The Three Arguments Given
in 1907
3. Excitation Energy and the
Emission Spectrum
4. Stark's Further Arguments of
1908
5. X Rays and the Light-Quantum
Hypothesis

5

Arthur Erich Haas

87

The First Application of
Quantum Theory to the Atom

(1910)

1. Atomic Models
2. Planck's Resonators
3. Arthur Erich Haas
4. The 1910 Theoretical Attempts
5. First Reactions to Haas's Ideas
6. Quantum-Theoretical Atomic
Models of 1911

6

Arnold Sommerfeld

103

Interactions between Electrons
and Molecules

(1910-1912)

1. Skeptical Reticence
2. A Change of Philosophy

3. Debye's Derivation of Planck's
Radiation Formula
4. Sommerfeld's First Quantum
Study
5. Planck's and Sommerfeld's
Hypotheses
6. Application and Criticism
7. Historical Effects

7

Walther Nernst
124

The Search Becomes Organized
(1910-1912)

1. Experiments on Quantum
Phenomena
2. Measurements of Specific Heat
3. Born and Kármán's and Debye's
Theories
4. The Quantization of Rotational
States
5. Preparation for the Solvay
Conference
6. The Solvay Conference of 1911

8

Niels Bohr
146

The Quantum Theory of the
Atom
(1912-1913)

1. The Introductory Work of 1912
2. The Atomic Energy Levels
3. The Reshaping of Stark's
Atomic Theory
4. Derivation of the Hydrogen
Spectrum

Further Reading
161
Name Index
163

The principle of continuity of all natural processes can be considered the godfather of the new physics which evolved in the 17th century. This is particularly evident in the development of the differential and integral calculus, whose spirit is also that of physics. The entire philosophy of Leibniz is imbued with the continuity principle in the broadest sense.

Leibniz clearly stated that the present holds the future in its lap and that any given state can be explained only by that which immediately preceded it. If this is disputed, then there will be gaps in the world which would topple the great principle of sufficient reason, forcing us to seek refuge in miracles or pure chance to explain natural phenomena.

During the course of the 18th and 19th centuries, the principle of continuity of all physical processes was implied by many scientists and philosophers, even though it was generally taken for granted and therefore not specifically formulated.

An act of desperation—this is what Max Planck later called his derivation of the radiation law: "Briefly summarized, what I did can be described as simply an act of desperation. By nature I am peacefully inclined and reject all doubtful adventures. But . . . a theoretical interpretation . . . had to be . . . found at any cost, no matter how high . . . . The two laws [of thermodynamics], it seems to me, must be upheld under all circumstances. For the rest, I was ready to sacrifice every one of my previous convictions about physical laws . . . ."

December 14, 1900, the date on which Planck presented his derivation of the law of black-body radiation to the Physikalische Gesellschaft in Berlin, is generally recognized as the birthday of the quantum theory. In fact, Planck, a reluctant revolutionary, had overthrown the principle of the continuity of physical processes by his introduction of the discrete energy levels of a linear oscillator.

In contrast to what is generally believed, Planck was hardly aware of the impact of his discovery around the turn of the century.

If Planck actually did not and could not have had the impression in 1900 of having overthrown the continuity principle, what then was the nature of his "act of desperation" in the derivation of the radiation formula and the "sacrifice of physical convictions"? The sacrifice made quite consciously by Planck was to renounce the purely axiomatic-thermodynamic conception of the second law of thermodynamics which he had embraced and to accept the atomistic and probabilistic theoretical interpretation which until then he had vigorously rejected. As late as 1896 he had allowed his assistant Ernst Zermelo to carry out a polemic with Boltzmann on this subject.

In order to derive the law of black-body radiation, Planck was now forced to use the Boltzmann method. The fact that in the course of his considerations he also had to write $\epsilon = h\nu$ did not present any great additional difficulties to him. "This was purely a formal assumption and I really did not give it much thought except that, no matter what the cost, I must bring about a positive result."

He had been familiar since May of 1899 with the physical constant h and had obtained its numerical value with great accuracy from radiation measurements. Thus, right from the beginning (and again later) the gradual formation of quantum theory can be characterized by the introduction of the physical constant h into physics; it is from this point of view that we would like to take up the genesis of the quantum theory and describe its development. By contrast, the discussions on the continuity principle only began in 1908–09 and even then formed only a small part of the many approaches made toward a solution of the quantum problem.

Inseparably linked to the efforts made to understand the nature of the quanta of energy or of action are the discoveries made and the understanding gained of a new range of physical phenomena. The history of quantum theory represents a typical chapter in the history of physics and is not simply an intellectual struggle for the recognition or rejection of certain ways of thinking.

While the quantum concept was initially the viewpoint of a minority, the opinions of physicists in general started to swing in its favor around 1910–11. When in March 1913 Niels Bohr was successful in solving the most important problem of physics of that time, the problem of atomic structure, by using Planck's quantum of action, a milestone in the development had been reached. We shall consider the genesis of quantum theory to extend to this point.

1

Max Planck
New Physical Constants and
the Radiation Law
(1899–1906)

1. The Radiation Equilibrium

After studying the papers of Rudolf Clausius, the young Max
Planck became convinced that, besides the energy principle, the
second law of thermodynamics also possesses a fundamental significance whose possibilities had by no means been fully ex-
ploited. The doctoral dissertation[1] presented by the 21-year old
Planck in Munich addressed itself to this matter. "The impression
made by this paper on the physicists of that time was equal to
zero," Max Planck reminisced later: "However, this experience
did not prevent me, completely convinced as I was of the significance of my work, from continuing my studies of entropy.
I considered entropy as the most important property of any
physical system along with energy. Since maximum entropy
identifies final equilibrium, all laws of physical and chemical
equilibrium can be derived from a knowledge of entropy. I carried
out this work in detail in the course of various studies during the
years following, initially with respect to aggregate changes of state
and then for gas mixtures and finally for solutions."[2]

This work brought no particular success to Planck since the
publications of Josiah Willard Gibbs had already appeared. On the
other hand, Planck "discovered new ground in the field of
radiated heat." Gustav Kirchhoff, in 1859–1860, had been the
first to emphatically point out the existence of a universal
function: "The quantity I (black-body emissivity) is a function
of wavelength and temperature. It is extremely important to
determine this function. Considerable difficulties block the path
to an experimental determination; nevertheless, there is reason to
hope that it can be established in this manner since it is undoubtedly a simple function like all those that do not depend on
the properties of specific bodies—at least those which have been
discovered up to this time."[3]

Thus, the significance of the intensity distribution of black-body radiation had already been pointed out in the derivation of Kirchhoff's laws, even though experimental and theoretical difficulties still prevented an early determination of this function. The sensitivity of thermopiles available for intensity measurements was much too low, and only after the invention of the bolometer by Langley could a solution of this problem be contemplated.*

"In my opinion, the determination of the function I is sufficiently important to endow a professorship for this purpose," remarked Friedrich Paschen.[4] Willy Wien was equally convinced of the significance of this problem. Pioneering efforts were carried out by Paschen and Wien; their close collaboration resulted in Wien's radiation law of 1896. The results of their cooperative effort were reported by Paschen to Kayser in the following words:

Furthermore, I am pleased to be able to inform you that I believe I have found the function I of Kirchhoff's law. From my observations, I have determined the emissivity function

$$I = \frac{c_1}{\lambda^\alpha} \cdot e^{-c_2/\lambda T} ,$$

from which it follows that $\lambda_{max}\ T$ = constant, and total radiation $= T^{a-1}$; c_1 and c_2 are constants, and α has the value 5.5, which is subject to possible adjustment after a more complete evaluation of the experimental data . . . . This equation was given added significance by the fact that W. Wien derived the same equation theoretically but with α = 5.0. When I sent him [my] equation, he wrote me, indicating that he had already arrived at the same result some time ago. We shall now jointly publish our underlying arguments for this equation; for it is quite probable that the true law has been discovered to a close approximation.[5]

Wien's radiation equation was verified by further measurements in the years following and accepted as an expression based on experience until the middle of the year 1900; only its derivation remained vulnerable.

*For an experimental solution of the problem of black-body radiation, see Kaysers Handbuch der Spektroskopie, vol. 2, Leipzig 1902.

When Planck's attention was drawn to this problem, he was also fascinated by the universal applicability of Kirchhoff's function: "This so-called normal energy distribution represents something absolute, and since the search for absolutes has always appeared to me to be the highest form of research, I applied myself vigorously to its solution."[6]

Following the example of Wien, Planck now examined the electromagnetic radiation that is enclosed within a cavity having ideally reflecting walls. The existing arbitrary distribution of radiation intensity over the various spectral regions then remains unchanged, since individual partial waves interpenetrate without interacting. Thus, "normal distribution" can only be attained by interaction with resonators that both absorb and emit radiation. Since the normal distribution is independent of the type of resonator, Planck assumed the simplest case of linear Hertzian oscillators (see p. 89).

With the use of this concept, the equilibrium state can be established by interaction between the oscillators and the radiation field. This interaction must therefore be examined in order to calculate the normal distribution. If the electromagnetic radiation energy per unit volume and frequency interval is designated by $u(\nu, T)$ and the mean energy of the Hertzian oscillators by $U(\nu, T)$, then according to Planck,

$$u(\nu, T) = \frac{8\pi \nu^2}{c^3} U. \tag{1}$$

Planck found equation (1) by equating the emission and absorption of a slightly damped Hertzian oscillator.[7] Problems arising from this equation by assuming discrete energy levels of the resonators were first pointed out by Einstein in 1905 (see p. 53).

"The significant aspects of this equation, which was indispensible to me, is that according to it the energy of the resonant oscillator depends only on radiation intensity and frequency ν but not on any of its other properties."[8]

2. Derivation of Wien's Equation

Planck announced the equation $u = 8\pi\nu^2 U/c^3$ on May 18, 1899, at a meeting of the Prussian Academy of Sciences. It is significant—as has been pointed out repeatedly[9,10]—that Planck did not immediately substitute for the energy of the linear oscillator U the following expression, which derives directly from the "classical" equipartition theorem:

$$U = k \cdot T. \tag{2}$$

This would have led Planck to the radiation equation

$$u = \frac{8\pi\nu^2}{c^3} \cdot kT \tag{3}$$

as early as 1899. It was actually first expressed by Rayleigh in June 1900.[11]

This leads immediately to the conclusion, perhaps quite startling at first glance, that Planck was unaware of the equipartition theorem. However, this becomes quite plausible if one examines the historical circumstances: on the one hand, Planck had hardly been familiar with the statistical viewpoints that led to the principle of equipartition, views which he specifically rejected until his "act of desperation" late in 1900. Furthermore, he reported in his Reminiscences[12] that the asymptotic behavior of radiation intensity $u(\nu, T)$ for $\lambda T \longrightarrow \infty$ was pointed out to him by Rubens (see p. 13) and ultimately led to Planck's radiation equation on October 19. However, if an equation as simple as (2) had already been available to Planck in 1899, he surely would have considered its consequences.* The derivation of (1) was therefore "no more than a preliminary step in an attack on the actual problem, whose magnitude now became all the more ominous . . . . Thus, I was forced to approach the problem from

---

*$U = k \cdot T$ leads directly to the entropy of the resonator $S = k \cdot \ln u$; this is not in agreement with Planck's definition of entropy of the resonator as stated in equation (9).

the other end, that is, by way of thermodynamics where I also felt more at home. Here, in fact, my previous studies of the second law of thermodynamics proved quite useful since it immediately occurred to me to determine a relationship between the energy of the resonator and its entropy rather than with its temperature."[13]

At that time, Wien's radiation equation

$$u = \frac{8\pi \nu^3 a'}{c^3} \cdot e^{-a\nu/T} \tag{4}$$

with its two parameters was accepted as valid. From this, Planck immediately derived the energy U of the resonator as

$$U = a'\nu e^{-a\nu/T}.$$

This yields, after solving for $1/T$, since $dS/dU = 1/T$,

$$\frac{dS}{dU} = -\frac{1}{a\nu} \ln \frac{U}{a\nu}. \tag{5}$$

After differentiation with respect to U, Planck obtained

$$\frac{d^2S}{dU^2} = -\frac{1}{a\nu} \cdot \frac{1}{U}. \tag{6}$$

Later, Planck introduced the symbol R for the reciprocal value of d^2S/dU^2, thus obtaining

$$R = -a\nu U. \tag{7}$$

He had now derived the following equation for entropy change:[13]

$$dS_t = \Delta U \, dU \cdot \frac{3}{5} \cdot \frac{d^2S}{dU^2}. \tag{8}$$

This describes a system whose entropy deviates from maximum entropy in that an individual resonator deviates by an amount ΔU from the equilibrium value U. The given entropy change occurs when the energy of the oscillator varies by dU. If ΔU and dU have differing signs, i.e., if the system returns to equilibrium, then dS_t must be positive; in other words, the function d$^2 S/dU^2$ must necessarily have a negative value. This requirement is satisfied by equation (6) which follows from Wien's radiation law.* "Since the entire problem deals with a universal physical law and since I was convinced, then as now, that physical laws tend to be simpler, the greater their generality . . . I believed for a time that I saw the basis for the entire energy-distribution law in the statement that the quantity R is proportional to the energy."[14]

Planck now believed that he could define the entropy of a resonator in such a way that R would be directly proportional to U. Thus, he wrote the following definition:

$$S = -\frac{U}{av} \ln \frac{U}{ea'v},$$ (9)

since from this immediately follows (5); and together with (1), Wien's radiation formula (4).

"I have repeatedly tried to change or generalize equation (9) for the electromagnetic entropy of a resonator in such a way that it still satisfies all theoretically sound electromagnetic and thermodynamic laws, but I was unsuccessful in this endeavor."[15]

This brought Planck to the following premature statement in May 1899: "I believe this must lead me to conclude that the definition of radiation entropy and therefore also Wien's energy-distribution law necessarily result from the application of the principle of the increase of entropy to the electromagnetic radiation theory and that therefore the limits of validity of this law, insofar as they exist at all, coincide with those of the second law of thermodynamics. Of course, this further increases the fundamental importance of additional experimental investigation of this law."[16]

We know that such measurements continued to be carried out

*The constant a which occurs in the radiation equation is positive and so is U.

intensively, but they were hardly influenced, at least initially, by Planck's perceptions.

3. The New Physical Constants

Planck had defined the entropy of a resonator by equation (9) "where a and a' designate two universal positive constants." Here, a' equals Planck's quantum of action h, a constant for which Planck stated a value of $a' = 6.885 \cdot 10^{-27}$ erg-sec in May 1899. In present-day terminology, $a = h/k$, where k is the Stefan-Boltzmann constant.

Planck recognized that he had discovered some important fundamental physical constants:

In this regard, it should be of some interest that use of the two constants a and a' which occur in the equation for radiative entropy offers the possibility of establishing units for length, mass, time, and temperature which are independent of specific bodies or materials and which necessarily maintain their meaning for all time and for all civilizations, even those which are extraterrestrial and nonhuman, constants which therefore can be called 'fundamental physical units of measurement.' Means for defining the four units of length, mass, time, and temperature are given by the above-mentioned constants a and a', by the value of the speed of light in a vacuum c, and by the gravitational constant f.[17]

The meeting of the Prussian Academy of Sciences at which Planck presented these ideas took place on May 18, 1899. Thus, if we consider those natural phenomena in which the physical constant $h = 6.27 \cdot 10^{-27}$ erg-sec plays a role (while the complementary quantity may also be written as $h = 0$) as the border between quantum theory and classical physics, then the birthday of quantum theory may well be placed as early as May 18, 1899.

In his subsequent considerations of the radiation equation, Planck also kept in mind the significant aspects of natural constants. On December 14, 1900, when he announced the derivation of what was now no longer Wien's but rather Planck's law—a date more generally known as the birthday of quantum theory—Planck also considered the more exact meaning of the natural constants k and h which in the meantime had become clear to him (see p. 20).

4. Planck's Radiation Equation

When Friedrich Paschen began his measurements, he was as yet
unaware that a black body could be realized by a radiation
cavity. Wien and Lummer pointed out this possibility in 1895 and
it was carried out by Lummer and Pringsheim in 1897. Lummer
and Pringsheim turned their attention to the verification of
the Stefan-Boltzmann law of total radiation, using measurements
of spectral energy distribution in a cavity. In so doing, they observed deviations from Wien's radiation law.* These deviations
from Wien's radiation equation soon become noticeable at
high temperatures and long wavelengths. Since the experiments
were based on Wien's equation (4), and radiation intensity was
measured at definite temperatures and wavelengths, this constituted a determination of the two constants in Wien's radiation
equation. The fact that Wien's equation was invalid was then
indicated by the increasing deviation of the constants for
$\lambda T \longrightarrow \infty$. In their attempt to attain larger values of λ and T,
Lummer and Pringsheim carried out experiments at 18μ and
$1772°K$ around the middle of the year 1900.

At that time [at the meetings of the Physical Society] there
were also considerable differences of opinion concerning the
laws of thermal radiation. The principal participants in these
debates were Lummer, Pringsheim, Jahnke, Thiesen, Kurlbaum,
and Rubens who proposed and discussed the most diverse equations. However, the final decision was brought about not by
Lummer's experiments but rather by the measurements of Rubens
and Kurlbaum which specifically demonstrated that at elevated
temperatures the intensity of monochromatic radiation is proportional to the temperature. I remember very well that Rubens
told me at that time: "Nevertheless, one thing is certain, the
intensity of monochromatic radiation has temperature as one
factor, and the other factor is an expression which remains finite
with infinitely increasing temperature."[12]

*It was clear to Planck that the weak point in his derivation of Wien's
radiation law lay in his definition of the entropy of the oscillator (9). In
March 1900 he apparently carried out a direct calculation of equation (9)
in the course of a renewed analysis and thus arrived at a further verification
of Wien's law. However, the rapidly progressing experiments showed that
Wien's law and thus Planck's calculation were untenable.

These remarks by Planck prove that Rubens had pointed out to him the limiting case $\lambda T \longrightarrow \infty$.* At the time he proposed his radiation equation on October 19, 1900 he was not aware of Rayleigh's radiation law, which had appeared in the June 1900 issue of the Philosophical Magazine.[†]

Since these results [of the energy measurements carried out at very long wavelengths by Kurlbaum and Rubens] became known to me several days before the meeting [of the German Physical Society on October 19, 1900] through verbal communications from the authors, I still had time to draw relevant conclusions before the meeting and use the information to calculate the entropy of a resonating oscillator. If, at high temperatures T, the radiation intensity u becomes proportional to temperature, then the latter is also proportional to the energy of the oscillator, according to $u = 8\pi\nu^2 U/c^3$. Thus;

$$U = C \cdot T, \tag{10}$$

from which, due to $dS/dU = 1/T$, and integrating:

$$S = \ln U. \tag{11}$$

Therefore,

$$\frac{d^2S}{dU^2} = -\frac{C}{U^2}. \tag{12}$$

Thus, this relation replaces, for large values of U, equation (6) which applies only to small values of U.[18]

If we assume, like Planck, that (6) and (12) represent the two limiting cases of the true physical law, then expression (12), which follows from Rubens' observations, is valid for large T and therefore for large U, since the latter increases with T. But in this case, (12) vanishes more rapidly than (6); thus, only the reciprocals of expressions (6) and (12) should be added. This reciprocal value of d^2S/dU^2 was called R by Planck.

*Experiments had already shown that Wien's equation was quite accurate for small values of λT (below about $2000°C \cdot \mu$).
†Rayleigh's law is equivalent to Rubens' statement, as reported by Planck, if viewed as a limiting expression for long wavelengths and elevated temperatures.

Two simple limits of the function R had thus been determined: at low energies it is directly proportional to energy, while at high energies it is proportional to the square of the energy. It therefore becomes obvious that, for the general case, the quantity R should be equated to the sum of a term of the first power and a term raised to the second power of energy so that at low energy the first term is dominant and at high energies the second term predominates. Now we had established the new radiation formula.[14]

The data left by Planck readily permits a reconstruction of his calculation. Starting with the sum of (6) and (12), which Planck wrote in the form [19]

$$\frac{d^2 S}{dU^2} = \frac{\alpha}{U(\beta + U)}, \tag{13}$$

he obtained by integration,

$$\frac{\beta}{\alpha} \cdot \frac{dS}{dU} = \ln U - \ln (\beta + U), \tag{14}$$

which by further integration leads to S,

$$\frac{S}{\alpha} = \frac{U}{\beta} \ln \frac{U}{\beta} - \left(1 + \frac{U}{\beta}\right) \ln \left(1 + \frac{U}{\beta}\right), \tag{15}$$

an expression which shows that β is proportional to ν in accordance with Wien's displacement law* $S = f(U/\nu)$. On the other hand, (14) immediately leads to

$$U = \frac{\beta}{e^{-\beta/\alpha T} - 1},$$

where $\beta = h \cdot \nu$ can now be substituted. Due to (13), it is known that a is negative and equals $-k$. This led Planck to formulate the equation

$$U = \frac{h\nu}{e^{h\nu/kT} - 1}. \tag{16}$$

---

*Of course, the constants α and β could also be determined by examining the limiting case $\lambda T \longrightarrow 0$ (Wien's radiation law).

This is Planck's expression of 1906.[20] It is the first time this quantum-theoretical expression for the mean energy of a linear oscillator is explicitly stated in the literature.

Equation (16), together with (1) rewritten for wavelengths, provides the radiation equation with two constants which Planck presented at the meeting of the German Physical Society in Berlin on October 19, 1900:

$$u = \frac{C\lambda^5}{e^{c/\lambda T} - 1}. \tag{17}$$

Next morning, on October 20, my colleague Rubens paid me a visit and told me that after the meeting, during the preceding night, he had carefully compared my equation with his measured data and found completely satisfactory agreement in all cases.[21]

While Lummer and Pringsheim did report a deviation in a letter dated October 24,[22] they quickly withdrew their objection, "since Pringsheim has told me that the apparent deviations were caused by an error in calculation."[21]

5. The Boltzmann Method

But there now remained the most crucial theoretical problem of finding a proper explanation for this law, an exceptionally difficult task since it involved a theoretical derivation of an expression for the entropy of an oscillator by integrating equation (14). That expression can be written in the form:

$$S = \frac{a'}{a} \left[\left(\frac{U}{a'\nu} + 1 \right) \ln \left(\frac{U}{a'\nu} + 1 \right) - \frac{U}{a'\nu} \ln \frac{U}{a'\nu} \right]. \tag{18}$$

In order to give physical meaning to this expression, it was necessary to consider the nature of entropy in an entirely new way which led beyond the field of electrodynamics.[23]

Apparently, Planck had not realized until then that it would be impossible to derive the radiation equation by using phenomenological thermodynamics:

For even if the new radiation formula should prove to be absolutely accurate, it would have only limited value if it could be viewed only as the fortunate guess of an interpolation formula. For this reason, from the day of its formulation on October 19, I devoted myself to the task of finding the real physical meaning of this equation. That question led me directly to a consideration of the relationship between entropy and probability, that is, to the ideas of Boltzmann.[24]

At another point, Planck states:

Until then I had paid no attention to the relationship between entropy and probability, in which I had little interest since every probability law permits exceptions; and at that time I assumed that the second law of thermodynamics was valid without exceptions.[23]

In 1896, at a time when his investigations of thermal radiation had already begun, Planck was still a determined opponent to an atomistic view of the second law of thermodynamics. He believed that he could show a conflict between the second law on the one hand and kinetic gas theory and the laws of classical mechanics on the other. Through his assistant, Ernst Zermelo, he had carried out a polemic with Boltzmann[25] in *Annalen der Physik*. Like certain other physicists with a strong theoretical-philosophical orientation, Planck at that time was apparently under the influence of the positivism of Mach and Ostwald.

Here the compelling force of reality,[26] which was exerted throughout the development of physics, came into play. Even though Planck abhorred the statistical view of thermodynamics, the laws of nature forced him—in his efforts to find a derivation "at any cost"—to take this seemingly unavoidable road: "But since no other path appeared to be open to me, I now tried the Boltzmann method and wrote a general expression for an arbitrary state of an arbitrary physical system

$$S = k \cdot \ln W, \tag{19}$$

where W is the corresponding calculated probability* of the state."[27] As Planck himself emphasized repeatedly, he was the first to introduce the relation $S = k \cdot \ln W$, while Boltzmann merely stated the proportionality between S and $\ln W$.[28]

*Looking back, Planck quite rightly states that W must be the "correctly calculated probability." As Einstein showed in 1909, Planck's calculation itself really does not deserve this designation (see p. 20).

In a previously unpublished letter of October 7, 1931,[29] Planck described his derivation of the radiation equation as "an act of desperation." The decisive step* was his reach for a statistical view of thermodynamics which he had previously rejected vehemently. The result is expressed in equation (19).

Planck now had to apply the previously distasteful Boltzmann method. This first required that energy be divided into atomic portions so that a counting process could then be used. It is quite possible that Planck at this point had in mind an example given by Boltzmann.

In the paper dating from 1877 from which Planck had quoted, Boltzmann treats the distribution of λ energy elements on n molecules: "Let us assume n molecules each of which can acquire the kinetic energy

$$0, \epsilon, 2\epsilon, 3\epsilon \cdots, p\epsilon;$$

these kinetic energies should be distributed in all possible ways among the n molecules provided, however, that the summation of the kinetic energy of all molecules remains constant, that is, equal to $\lambda\epsilon = L$."[30]

This is equivalent to Planck's distribution of energy U_N over N resonators:

It is now important to determine the probability W that N resonators have a total vibrational energy U_N. This requires that U_N be viewed not as a constant quantity which can be infinitely subdivided but rather as a discrete quantity composed of a number of finite and identical parts.†

If such a part is called an energy element ϵ, then the following expression can be written:

$$U_N = P_\epsilon, \tag{20}$$

---

*The desperate step was by no means the relation $\epsilon = h \cdot \nu$ for the energy levels of the oscillator: "That energy is forced, at the outset, to remain together in certain quanta . . . was purely a formal assumption and I really did not give it much thought" (see p. 24).

†If U_N is infinitely divisible, then the distribution on N resonators can be made in an infinite number of ways.

where P is an integer, generally quite large, while the value of ϵ must still be determined.[31]

Like Planck, Boltzmann in his paper of 1877 had divided the "actually" continuously variable energy (i.e., velocity) into discrete portions: "If we consider that the infinite in nature is never anything but a transition to a limit, then the infinitely variable velocities which each molecule is capable of assuming can only be viewed as the limiting case which occurs when each molecule can assume more and more velocities."[32]

It is undoubtedly true that Planck initially suspected that such a transition to a limit could also be carried out in his calculation. He now determined the number of possible distributions of P energy elements on N oscillators which, like Boltzmann, he called complexions. According to the rules of combinatorial analysis, these are

$$W = (N + P - 1)!/(N - 1)!P!. \tag{21}$$

This is the distribution of P independent but undifferentiated quanta over N differentiated oscillators. This method of counting was clarified by Paul Ehrenfest[33] and Ladislas Natanson.[34] The former's significant contributions deserve fuller treatment than can be accorded here. The reader is referred to Klein's recently published Ehrenfest biography (see p. 161).

Planck now neglects unity in the numerator and the denominator (since N is a very large number) and then rewrites the resulting expression, using of Stirling's approximation $N! = (N/e)^N$, yielding

$$W = (N + P)^{N+P}/N^N P^P. \tag{22}$$

This expression is now substituted into the equation $S_N = k \cdot \ln W$ which, due to $S_N = N \cdot S$, leads directly to the following expression for the entropy of an individual resonator.

$$S = k \left[\left(\frac{U}{\epsilon} + 1 \right) \ln \left(\frac{U}{\epsilon} + 1 \right) - \frac{U}{\epsilon} \ln \frac{U}{\epsilon} \right]. \tag{23}$$

Up to this point, the magnitude of the energy element ϵ is completely arbitrary. But now Planck must compare this equation with the equation

$$S = \frac{a'}{a}\left[\left(\frac{U}{a'\nu} + 1\right)\ln\left(\frac{U}{a'\nu} + 1\right) - \frac{U}{a'\nu}\ln\frac{U}{a'\nu}\right],\qquad(18)$$

which must be regarded as an expression of the experimental results since it is equivalent to Planck's radiation equation.

The similarity between expressions (23) and (18) is obvious. Thus, there only remain those steps that are necessary to make these equations identical. This is achieved by setting $k = a'/a$ and by designating the magnitude of the energy element as $\epsilon = a'\nu$. The constant a', which is independent of oscillator characteristics, I designated by h and since it has the dimensions of a product of energy and time, I called it the elementary quantum of action or element of action in contrast with the energy element $h\nu$.[35]

Thus, Planck wrote down the energy element (that is, the difference between energy levels, of the oscillator) as

$$\epsilon = h\nu.\qquad(24)$$

The constant $h = 6.55 \cdot 10^{-27}$ erg-sec had already been familiar to him since May 1899 (even then he described h as a physical constant). This undoubtedly contributed to the fact that he regarded the relation (24), though not completely understood, as definitely established.*

This information from Planck indicates agreement with the historical view[36, 37] that he did in fact derive his radiation equation in the manner described. A few remarks should be added.

In deriving the decisive formula (21), Planck simply counted the total possible number of states and, by definition, set the total number equal to the probability. As early as 1899 (when he still accepted the Wien radiation equation as correct), Planck had

---

*Planck was therefore not compelled to let h converge to zero. By contrast, James Jeans demanded a transition to the limit $h \longrightarrow 0$.

already solved—or rather put off—some of the immense difficulties which faced him by simply determining the value of the entropy of the oscillator by definition.

A similar procedure was used in the derivation of Planck's radiation equation:

In my opinion this determination $S = k \cdot \ln W$, is basically a definition of the so-called probability W; for the presuppositions on which the electromagnetic theory of radiation is based provide no indication that would permit us to speak of such a probability in a definite sense."[31]

In 1909 and again in 1911, Albert Einstein pointed with laudable clarity to this weak point in Planck's derivation:

The manner in which Mr. Planck uses Boltzmann's equation is rather strange to me in that a probability of state W is introduced without a physical definition of this quantity. If one accepts this, then Boltzmann's equation simply has no physical meaning.[38]

On the other hand, Planck did not believe as late as 1911 "that there exists a definition of probability that is completely general and applicable outside of classical dynamics and that permits calculation of the probability of any arbitrary state."[39]

In fact, the problem of quantum statistics could only be resolved completely after quantum mechanics had been developed; the first contributions (aside from Planck and Einstein) were made by H. A. Lorentz, Ladislas Natanson, and Paul Ehrenfest.

6. The Meaning of the Natural Constants

No later than May 1899 did Planck begin to consider the meaning of the two physical constants k and h. By the middle of 1902 he had recognized that the first of these, k, depended on the definition of temperature. While considering the equation $S = k \ln W$, Planck stated:

The proportionality constant k depends on the units in which temperature is measured. If the temperature of a gas is directly set equal to the mean energy of an atom, then $k = 2/3$. But since the unit of temperature is determined by the conventional definition that the separation between the boiling point and the freezing point of water equals 100, the constant k has the dimensions of energy divided by temperature.[40]

The constant k therefore does not have the significance of a fundamental physical constant that might lend support to the search, then as now, for a final theory of physics as Planck had envisaged for a while. For in 1899, Planck had wanted to make k part of the foundation for a natural system of units.[41] Nevertheless, the Boltzmann constant k (which, historically, might be called "Planck's constant" just as readily as the quantum of action h) played an important role in the development of atomic theory during the first years of this century.

In the relation $R = k \cdot N_0$, R is the gas constant, already then well known, and N_0 is the number of molecules per mole which is still called Loschmidt's number in the German literature (and elsewhere is known as Avogadro's number), $N_0 = 6.02 \cdot 10^{23}$. Loschmidt had never mentioned such a number, much less given it a value. He had limited himself (1865) to providing an approximation for the size of a molecule, giving the molecular diameter a magnitude of about 10^{-8} cm.

From his theory of thermal radiation, Planck was able to arrive at a value $R/k = 6.175 \cdot 10^{23}$ which was quite accurate, considering that N_0 first appeared in the physics literature during the last years of the 19th century. The electrical charge carried by a gram equivalent of monovalent ions was known from electrolytic theory (it is now often designated as Faraday's constant). From this, Planck could immediately determine the elementary electrical quantum using the number N_0. He found a value of $4.69 \cdot 10^{-10}$ esu. At that time, Planck was only familiar with the measurements of Franz Richarz who had found $1.29 \cdot 10^{-10}$ and of J. J. Thomson who gave a value[42] of $6.5 \cdot 10^{-10}$. The present value is $4.8 \cdot 10^{-10}$ esu.

I could derive some satisfaction from this result. But matters were viewed quite differently by other physicists. Such a calculation of an elementary electrical quantum from measurements of thermal radiation was not even given serious consideration in some quarters. But I did not allow myself to become disturbed by such a lack of confidence in my constant k. Nevertheless, I only became completely certain on learning that Ernest Rutherford and Hans Geiger obtained a value of $4.65 \cdot 10^{-10}$ by counting alpha particles.

These investigations were carried out in 1908 and 1909; at about the same time Erich Regener obtained similar results.

It can be viewed as a certain irony of history that Max Planck, who only five years earlier had carried out a polemic against the atomistic views of Boltzmann through his assistant Ernst Zermelo, should open new insights into the atomic nature of matter and electrical charges. Besides mentioning "gram molecules," he specifically talked of "real molecules." Thus, as early as 1900, he had already fully recognized the physical significance of the constant k in his radiation law.

Despite strenuous efforts, Planck did not succeed in developing an interpretation of the second constant, h. The clarification process, as we know today, comprises the history of quantum theory and required almost three decades.

It should be noted that Max Planck occasionally did consider a relationship between the atomic nature of electricity and the quantum of action h. On July 6, 1905, he wrote the following note to Paul Ehrenfest (see page 60):

It seems to me not completely impossible that there can be a connection between this assumption (the existence of an elementary quantity quantum of electricity) and the existence of an elementary quantum h, since h has the same order of magnitude as e^2/c (where e is the elementary quantity of electricity in electrostatic units and c is the speed of light)."[43]

In 1909, Einstein attempted to develop this idea further.

7. "One of the Greatest Discoveries since Newton"
Driven by the necessity to find a derivation for the radiation law, Planck had converted from an opponent to a proponent of atomic theory* and of the statistical view of the second law of thermodynamics to which Boltzmann subscribed.

---

*But Planck's fundamental acceptance of the atomic view by no means meant that he would have shown understanding for and agreement with definite models of atoms. On the contrary, he held considerable reservations in this matter.

"I derived great satisfaction," reported Planck, "from Ludwig Boltzmann's letter responding to the paper I had sent him, wherein he expressed his interest and his fundamental agreement with the direction that my ideas had taken."[44] It would be of great historical interest to find this letter of Boltzmann; but apparently it was destroyed during the war, together with Planck's other scientific papers. Fortunately, there does exist another (though much later) letter addressed to Robert Williams Wood, in which Planck describes in detail the psychological aspects of his approach. Because of its special significance this letter, to which we have already referred several times, will here be reproduced in full:[29]

My Dear Colleague,

You recently expressed the wish, after our fine dinner in Trinity Hall, that I should describe from a psychological viewpoint the considerations which had led me to propose the hypothesis of energy quanta. I shall attempt herewith to respond to your wish.

Briefly summarized, what I did can be described as simply an act of desperation. By nature I am peacefully inclined and reject all doubtful adventures. But by then I had been wrestling unsuccessfully for six years (since 1894) with the problem of equilibrium between radiation and matter and I knew that this problem was of fundamental importance to physics; I also knew the formula that expresses the energy distribution in normal spectra. A theoretical interpretation therefore *had* to be found at any cost, no matter how high. It was clear to me that classical physics could offer no solution to this problem and would have meant that all energy would eventually transfer from matter into radiation.* In order to prevent this, a new constant is required to assure that energy does not disintegrate. But the only way to recognize how this can be done is to start from a definite point of view. This approach was opened to me by maintaining the two laws of thermodynamics. The two laws, it seems to me, must be upheld under all circumstances. For the rest, I was ready to sacrifice every one of my previous convictions about physical laws. Boltzmann had explained how thermodynamic equilibrium

---

*This statement sounds as though Planck in 1900 already had a clear idea of the consequences of Rayleigh's classical radiation formula, equation (3). But this was obviously not the case. It is probable that Planck attained this insight toward the end of 1901 as a result of a paper by James Jeans[45] in which this thought was expressed.

is established by means of a statistical equilibrium, and if such an approach is applied to the equilibrium between matter and radiation, one finds that the continuous loss of energy into radiation can be prevented by assuming that energy is forced, at the outset, to remain together in certain quanta. This was purely a formal assumption and I really did not give it much thought except that, no matter what the cost, I must bring about a positive result.

I hope that this discussion is a satisfactory response to your inquiry. In addition, I am sending you as printed matter the English version of my Nobel lecture on the same topic. I cherish the memory of my pleasant days in Cambridge and the fellowship with our colleagues.

With kind regards,
very truly yours
M. Planck.

The contents of this letter bears out our own view of the historical development: An "act of desperation"—that was the introduction of the relationship between entropy and probability. The phrase "and I really did not give it much thought" refers to the introduction of energy elements.

Nevertheless, Planck had clearly realized that without these elements a derivation of the radiation law would not be possible. In his lecture of December 14, 1900, he had emphasized that "we take the view—and this is the most significant point of the whole calculation—that [the energy of an oscillator] E is composed of a quite definite number of finite and equal parts and for this we use the natural constant $h = 6.55 \ 10^{-27}$ erg-sec."[46]

During long walks in the Grunewald, Planck is said to have talked to his favorite son Erwin some time near the turn of the century about having succeeded in making "one of the greatest discoveries in physics since Newton." This story is related by Werner Heisenberg. But Heisenberg heard this not from Erwin or from Max Planck but only through third parties. There are two versions of Heisenberg's report, both of which will be given here:

Recalling those days, his son Erwin Planck said that while he and his father were walking in the Grunewald, the latter excitedly discoursed during the entire walk about the results of his investigations. He is said to have told his son something along these lines: Either what I have found out now is complete nonsense or it might be one of the greatest discoveries in physics since Newton.[47]

At a later time, Planck's son is said to have told of a long walk through the Grunewald with his father when he was a child while his father talked about his new ideas. Planck explained to his son that he had the feeling of either having made discovery of the first magnitude, possibly comparable to the discoveries of Newton, or else that he would prove to be completely in error.[48]

Are we dealing here with something more than a legend? Is it possible that Planck in order to be understood by his son Erwin, who was then seven years old, had made use of superlatives which at other times were foreign to him?

The history of physics is rich in legends. Just like Galileo's "Eppur si muove," Planck's statement of the "greatest discovery since Newton" (stated as early as 1900 or 1901) appears to us like words placed in the mouth of the hero by posterity. Despite their questionable origin, such reports are typical for the time in which they developed. In the case of Planck, they reflect the veneration of a subsequent generation of physicists for Planck and recognition of his great achievement.

In order to carry out the derivation of the radiation formula (which required a total of six years), a "very clear, logical head" was required. Planck was well qualified in this respect as had already been pointed out by his teachers at the Maximilian High School in Munich.* But in addition to this, an unusually systematic and conscientious personality was required for this task.

The extent to which Planck was conscientious and orderly is almost legendary. This was demonstrated even in his favorite spare-time activities, mountain climbing and piano playing, delightfully described by Heisenberg.[51]

If in fact, despite our doubts, Planck had given the status of "one of the greatest discoveries since Newton" to his derivation

*From a report card: "Rightfully the favorite of his teachers and his colleagues. He is the youngest in his class and despite being somewhat childlike has a very clear, logical head. Shows considerable promise."[49]

around the turn of the century*, it would be our belief, in agreement with Planck's letter to Wood, that this remark of Planck does not refer to the introduction of elements of energy into physics but rather in a more general way to the discovery of two fundamental natural constants. Of course, Planck's expression $\epsilon = h \cdot \nu$ represents a complete break with classical physics, one of whose cornerstones is the continuity principle. Planck realized the radical consequences of his formula only much later, and for many years he made attempts to harmonize the quantum of action with classical theory.

This also agrees with a statement of Fritz Reiche who heard Planck's lecture on thermal radiation in 1904 and later recalled the general impression that Planck had made on his students. According to Reiche, Planck did not believe that he had made a complete break with the past. He gave the impression that it was impossible to speak of a "counted" probability without a subdivision of energy.[52]

References

1. Max Planck, Physikalische Abhandlungen und Vorträge, Braunschweig 1958, vol. 1, pp. 1–61.

2. Ibid., vol. 3, p. 257.

3. Gustav Kirchhoff, *Poggendorff's Annalen der Physik,* vol. 109, 1860, p. 292.

4. Friedrich Paschen, letter to Heinrich Kayser, Staatsbibliothek Preussischer Kulturbesitz. Sign. Darmst. F 1e 1893, February 8, 1898.

*By 1907, Planck was certainly no longer convinced that he had made one of the greatest discoveries since Newton. In a letter addressed July 6, 1907, to Einstein in which Planck emphasizes the importance of agreement among the adherents to the principle of relativity, there appears the following sentence following a discussion of questions concerning black-body radiation: "More urgent than this question, which is after all quite in the background, is the matter of the acceptance of the relativity principle."[50] Only after the work of Einstein and others, the beginning of dialogue with his colleagues, together with the fact that the special theory of relativity had already become accepted by leading physicists as early as 1908, did quantum problems again become matters of great interest to Planck.

5. Ibid., letter, June 4, 1896. See also *Wiedemann's Annalen der Physik,* vol. 58, 1896, pp. 662–669.

6. Ref. 1, vol. 3, p. 389.

7. Ibid., vol. 1, p. 575.

8. Ibid., vol. 3, p. 260.

9. Martin J. Klein, *Archive for History of Exact Sciences,* vol. 1, 1962, pp. 459–479.

10. Léon Rosenfeld, *Osiris,* vol. 2, 1936, pp. 149–196.

11. Lord Rayleigh, *Philosophical Magazine,* Ser. 5, vol. 49, 1900, pp. 539–540.

12. Ref. 1, vol. 3, p. 407.

13. Ibid., vol. 1, p. 679.

14. Ibid., vol. 3, p. 124.

15. Ibid., vol. 1, p. 596.

16. Ibid., vol. 1, p. 597.

17. Ibid., vol. 1, p. 666.

18. Ibid., vol. 3, p. 262.

19. Ibid., vol. 1, p. 688.

20. Max Planck, Vorlesungen über die Theorie der Wärmestrahlung, Leipzig 1906, p. 127.

21. Ref. 1, vol. 3, p. 263.

22. Otto Lummer and Ernst Pringsheim, *Naturwissenschaften,* vol. 29, 1941, p. 137.

23. Ref. 1, vol. 3, p. 264.

24. Ibid., vol. 3, p. 125.

25. Ernst Zermelo, *Wiedemann's Annalen der Physik,* vol. 57, 1896, pp. 485–494.

26. Friedrich Hund, Geschichte der Quantentheorie, Mannheim 1967, p. 9.

27. Ref. 1, vol. 3, p. 265.

28. Arnold Sommerfeld, Thermodynamik und Statistik, 2nd edition, Leipzig 1962, p. 181.

29. Max Planck, letter to Robert Williams Wood, Sources for History of Quantum Physics, Mf. 66, 5, October 7, 1931.

30. Ludwig Boltzmann, Wissenschaftliche Abhandlungen, Leipzig 1909, vol. 2, p. 168.

31. Ref. 1, vol. 1, p. 720.

32. Ref. 30, vol. 2, p. 167.

33. Paul Ehrenfest, Collected Scientific Papers, Martin J. Klein, ed., Amsterdam 1959.

34. Ladislas Natanson, *Physikalische Zeitschrift*, vol. 12, 1911, pp. 659–666.

35. Ref. 1, vol. 3, p. 266.

36. Léon Rosenfeld, Max Planck-Festschrift, Berlin 1958, p. 204.

37. Ref. 9, p. 470.

38. Die Theorie der Strahlung und der Quanten, Arnold Eucken, ed. [Transactions of the 1st Solvay-Congress], Halle 1913, p. 95.

39. Ibid., p. 359.

40. Ref. 1, vol. 1, p. 735.

41. Josef Loschmidt, *Sitzungsberichte der math.-naturw. Classe der Kaiserl. Akademie der Wissenschaften*, vol. 52, 1866, p. 404.

42. Ref. 1, vol. 1, p. 730.

43. Max Planck, letter to Paul Ehrenfest, Rijksmuseum Leiden, Ehrenfest Collection, July 6, 1905.

44. Ref. 1, vol. 3, p. 126.

45. James Jeans, *Philosophical Transactions of the Royal Society,* Ser. A, vol. 196, 1901, pp. 397–430.

46. Ref. 1, vol. 1, p. 700.

47. Werner Heisenberg, Stimmen aus dem Maxgymnasium, no. 6, München 1958, p. 9.

48. Werner Heisenberg, Physik und Philosophie, Berlin 1959, p. 16. See also Max Born, *Max Karl Ernst Ludwig Planck*, 1858-1947, Obituary Notices of the Royal Society, vol. 6, 1948, p. 170.

49. Bernhard Winterstetter, Stimmen aus dem Maxgymnasium, no. 6, München 1958, p. 2.

50. Max Planck, letter to Albert Einstein. Einstein Archives, Princeton, N.J., July 6, 1907.

51. Ref. 47, p. 14.

52. Interviews with the Sources for the History of Quantum Physics. Manuscripts in the Niels Bohr Archives, Copenhagen (oral records): Fritz Reiche.

H. A. Lorentz   Discussion of the Radiation
                                     Problem
                                     (1903–1910)

1. The Reception of Planck's Radiation Theory

Within a few weeks after Planck's radiation equation was first
announced on October 19, 1900, it was readily accepted by
physicists involved in radiation-intensity measurements as being
experimentally well verified. Despite occasional doubts, this response rapidly became stronger.

At that time, Heinrich Kayser was completing the second volume of his "Handbook of Spectroscopy" for which he drew on
the assistance of Friedrich Paschen in questions concerning the
radiation of solid bodies with which he was not closely familiar.
On December 30, 1900, Paschen wrote to Kayser:

It seems that the matter of the radiation law will soon be fully
resolved in favor of the modification of Wien's equation that was
recently discovered by Planck. Two or three papers remain to be
published on this subject, but these should appear within a few
weeks. It would be advisable to have this matter covered as fully
as possible in your work. Recent discussions on this subject were
held mostly in private and I do not know whether you are familiar
with the present status. I for one do not know whether Planck has
already succeeded in providing his new law with a reasonably
sound explanation. I would want particularly to wait for clarification of this point.[1]

After Kayser had sent his draft of the material dealing with
black-body radiation to Paschen, the latter made the following
remarks concerning Kayser's manuscript on January 3, 1901:

What is still missing in this article is the following: Planck's new
law, which is identical with Wien's law for values of λT up to
$3000\mu \cdot {}^{\circ}C$, then approximately follows Thiesen's formula for
larger values of λT, and ultimately undergoes a transition to
Rayleigh's equation for very large values of λT, has been completely verified by all the investigations of Wanner and myself,
the only exception being my observations on energy curves at the
greatest wavelengths and the most elevated temperatures. But
even here, new experiments carried out differently have been
found to agree with Planck's law. The observations of Lummer
and Pringsheim, at least at longer wavelengths and high temperatures, follow this same law. Perhaps they will also find that same
agreement at smaller wavelengths and lower temperatures (where

there still exist considerable differences). In that case we would have complete agreement between all investigations and Planck's law.[2]

When the second volume of Kayser's handbook appeared in 1902, it was the first monograph in which Planck's radiation equation and its derivation were included. Planck's theory is only briefly mentioned: "Planck later derived his formula in a somewhat different form, introducing two new constants:

$$u_\nu d\nu = \frac{8\pi h \nu^2}{c^3} \cdot \frac{d\nu}{e^{h\nu/kT} - 1} \quad ,$$

where u_ν is the density of the radiated energy in the spectral region from ν to $\nu + d\nu$. The equation is explained theoretically, and it is shown that the constants h and k are natural constants related to the physical dimensions of a molecule and the size of the electron."[3]

None of the other textbooks and handbooks of physics which appeared in subsequent years go beyond the description given by Kayser. Where Planck's radiation formula is mentioned at all, there often is no reference to the natural constants which enter into this equation. Thus, the 1906 edition of Lommel's textbook on experimental physics mentions only the Stefan-Boltzmann law and Wien's displacement law.[4] In the textbook by Chwolson (1904) there appears the statement: "At the present time, the formula of Lummer and Jahnke and that of Planck must be viewed as the optimal expressions of the radiation law for black bodies. Some preference should be given to the latter, since the former has a rather empirical character."[5] Chwolson states Planck's formula and the values of the two constants h and k but he gives no indication that these are fundamental natural constants.

Eduard Riecke does discuss the significance of the constant k (but not of h) in the 1908 edition of his textbook: "Planck has applied electromagnetic theory to the problems of radiation . . . . One of the significant results of Planck's theory is that it leads in a novel way to the number of molecules actually contained in a gram mole. This is based on a property of the constant k."[6]

In the handbook of physics published by A. Winkelmann, which appeared in 1906, there is no mention of the derivation for Planck's formula; this is typical for all other presentations of that time. Here only Planck's expression for the entropy of an oscillator is given: "Among these expressions describing entropy, is a simpler one introduced by Planck in a subsequent study,

$$\Phi = k \left[\left(1 + \frac{U}{h\nu}\right) \ln \left(1 + \frac{U}{h\nu}\right) - \frac{U}{h\nu} \ln \frac{U}{h\nu} \right],$$

which also contains two constants, k and h."[7] The statement is then made that Planck's radiation formula follows from this entropy value. The subsequent text also completely disregards Planck's most significant achievements of 1900 and simply reflects the status as of May 1899:

Planck points out that the two constants k and h, together with the velocity of light and the absolute gravitational constant, furnish four constants of varying dimensions which are independent of our special conditions on earth and which therefore might also serve as independent units for measuring lengths, mass, time, and temperature.

This clearly shows that the community of physicists of that time had not become aware that something revolutionary was launched with Planck's radiation theory. During the first years of the 20th century, only Lord Rayleigh, James Jeans, and H. A. Lorentz* entered into serious discussions of the essential problems which now arose.

As early as June 1900, when Planck still believed in the validity of Wien's radiation law, Lord Rayleigh had published a short paper on the black-body radiation law. In it, Rayleigh considers Wien's radiation law for a given fixed frequency in the limiting case $T \longrightarrow \infty$. The radiation intensity then converges to a fixed limit independent of T. He noted earlier than Planck that this re-

*Albert Einstein followed in 1905. He was the first to demonstrate true comprehension of the basic questions to which Planck addressed himself.

sult contradicts Rubens's measurements: "It is true that for visible rays the limit is out of range. But if we take $\lambda = 60\mu$ (according to the remarkable researches of Rubens) for the rays selected by reflection of surfaces of Sylvin, we see that for temperatures over $1000°$ absolute there would be but little further increase of radiation."[8]

Since these experiments went contrary to Wien's law, Rayleigh attempted to derive a new radiation law; he proposed what later became known as the Rayleigh-Jeans law which appeared to him "more probable a priori."[8]

In his *Theory of Sound* Rayleigh had already addressed himself to elastic resonant vibrations in detail as early as 1877. He now counted the electromagnetic resonant vibrations $Z(v)$ in a radiation cavity and found

$$Z(v) \sim v^2.$$

Applying the classical mechanical equipartition theorem, with whose limitations he was quite familiar, he stated: "According to this doctrine every mode of vibration should be alike favoured; and although for some reason not yet explained the doctrine fails in general, it seems possible that it may apply to the graver modes."[8] Thus, Rayleigh obtained

$$u \sim v^2 T. \tag{1}$$

Apparently Rayleigh noted that the expression for total radiation diverged according to his law, because he introduced a "cut-off factor" $\exp(-c_2/\lambda T)$ and wrote

$$u \sim v^2 T e^{-c_2/\lambda T} \tag{2}$$

In 1905, he stated the proportionality constant for equation (1) but erred[9] in this by a factor of 8, but this was immediately corrected by James Jeans.[10] For this reason, the formula

$$u(v,T) = \frac{8\pi v^2}{c^3} k T \tag{3}$$

is today generally called the Rayleigh-Jeans radiation law. This is the well-known and strictly classical expression for the spectral energy distribution of black-body radiation.

In 1900, Rayleigh explicitly considered the laws (1) and (2) as being applicable only to the limiting case $\lambda T \longrightarrow \infty$. He stated that (1) "may be the proper from when λT is great".[11] In the 1902 edition of his Collected Papers he again emphasized the validity of his formula for $\lambda T \longrightarrow \infty$: "Very shortly afterwards the anticipation . . . was confirmed by the important researches of Rubens and Kurlbaum . . . , who operated with exceptionally long waves. The formula of Planck, given about the same time, seems best to meet the observations. According to this modification of Wien's formula, $\exp{(-c_2/\lambda T)}$ in (2) is replaced by $1/\exp{(c_2/\lambda T)} - 1$. When λT is great, this becomes $\lambda T/c_2$ and the complete expression reduces to (3)."[11] Later he leaned toward the views of James Jeans who postulated a general validity of the law. Jeans had recognized very well that the total energy of radiation diverged according to his formula:

If an interaction between aether and matter exists, no matter how small this interaction may be, the complete dynamical system will consist of the molecules of the gas, together with the aether, and must therefore be regarded as a system possessing an infinite number of degrees of freedom. Applying Boltzmann's Theorem to this system, we are merely led to the conclusion that no steady state is possible until all the energy of the gas has been dissipated by radiation into the aether.[12]

Jeans saw the solution to the problem in that the total transition of all energy into radiation as required by the theory takes place very slowly so that the measured intensity distribution does not agree with the equilibrium state:

We can now trace the course of events when one or more masses of gas are left to themselves in undisturbed aether. At first we may suppose that the total energy is . . . that of the principal degrees of freedom . . . a transfer of energy is taking place between the principal degrees of freedom of the molecules, and the vibrations of low frequency in the aether. This . . . endows the aether with a small amount of energy . . . After this, a third transfer of energy begins to show itself, but the time required for this must be measured in millions or billions of years unless the gas is very hot . . . if the whole system is enclosed by an ideal perfectly reflecting boundary, the energy accumulates in the aether . . . .[13]

Planck could not agree with such an assumption; it must have appeared to him as completely out of the question. All previous views, starting with Kirchhoff's law and continuing with the Stefan-Boltzmann total radiation law and Wien's displacement law, were based on an assumption of thermodynamic equilibrium. Similarly, it had never been shown experimentally that the measured intensity distribution demonstrates systematic variations with time.

Planck searched for a way out, stating "that the difficulties under discussion are only the result of an unjustified application of the principle of equipartition of energy distribution to all independent variables of state."[14] Planck recognized in 1906 that the "elementary regions" in the phase plane cannot be chosen arbitrarily small but that their magnitude must be equal to h. In 1911 Friedrich Hasenöhrl clarified this statement (see p. 100).

2. Struggles With the Problem

Hendrik Antoon Lorentz had extended Maxwell's theory in the 1890's into his famous electron theory, taking into consideration the atomic structure of matter, and at the turn of the century he furnished equally significant contributions to the electrodynamics of moving bodies. Soon after the development of the radiation theory of Planck and Rayleigh-Jeans, H. A. Lorentz had also "wrestled continuously with this problem."[15]

As the founder of electron theory, Lorentz first attempted to solve the problem from that direction. Paul Drude and Eduard Riecke had successfully treated electrical conductivity in metals using the concept that free electrons in the metal undergo thermal agitation. When a potential is applied, these electrons move preferentially in a single direction and at intervals undergo collisions with atoms. Thus, the Drude theory indicates a temperature dependence of conductivity, $\sigma \sim 1/T$.

Planck had treated the optical problem of metallic reflection in the same manner. In 1903, Hagen and Rubens carried out penetrating investigations using infrared residual radiation which was filtered out by multiple reflection from certain crystals (such as rock salt or fluorite) from a wide frequency spectrum.[16] Metallic reflection occurs in these narrow frequency bands, and the reflectivity R depends on wavelengths and electrical conductivity according to the phenomenological Lorentz-Maxwell theory[17]

$$1 - R = 1/\sqrt{\sigma\lambda} \cdot$$

This relation is also subject to the Drude theory. Its dependence on wavelength and on temperature was experimentally verified by Hagen and Rubens, while satisfactory agreement between the optically measured conductivity σ and the values known from electrical resistance was also demonstrated.

Planck, Hagen, and Rubens had successfully traced the absorption of electromagnetic radiation* to the effect of free electrons. Here, Lorentz saw a point of attack: according to Kirchhoff's law, the emissivity is dependent on absorptivity so that both must be based on the same mechanism. Thus, it should be possible to relate this to the Kirchhoff function and to black-body radiation. Using the relationship

$$u \, d\nu = \lambda^{-5} \, f(\lambda, T) \, d\lambda$$

for the electromagnetic energy density of radiation, Lorentz in 1903 obtained

$$f(\lambda, T) = 8\pi\lambda k \, T,$$

the Rayleigh-Jeans radiation law.[18] He well knew that the measurements of Hagen and Rubens applied only to long wavelengths. He now attempted to extend these considerations to short waves in order to obtain a radiation formula with general

*The reflectivity R is defined as the ratio of reflected to absorbed radiation energy.

validity. He discussed these efforts in a letter written to Willy Wien on June 6, 1908:

It seemed to me that it should be possible to derive the radiation laws, and in particular the physical significance of the constant $\lambda_m T$ by considering the mechanism of the phenomena. My thoughts along these lines were reinforced by the fact that satisfactory results are obtained for long wavelengths when the emission and absorption of a metal are calculated under the assumption of only free electrons moving in the interstices between metallic atoms. This viewpoint was suggested by the success of Drude's electron theory on the one hand and by the measurements of Hagen and Rubens on the other. For the latter have shown that it is possible to derive the absorptivity of a metal for long wavelengths from its conductivity: since electrical conductivity is definitely based on the movement of free electrons, it must obviously be possible to explain absorption in the same way. The same should therefore also be possible for emission, since absorptivity and emissivity would not show similar characteristics if there were not some close relationship between the manner in which radiation is absorbed and in which it is emitted.

It does not seem far-fetched to say that the radiation function E/A will be obtained correctly if E and A are calculated for any arbitrary *imagined* body—at least if the assumptions made about this body are not too extravagant. If this is correct, then we can base our considerations on a body having an infinite number of resonators or else, with just as much justification, a body containing only free electrons. I had therefore hoped also to carry out the determination of the function E/A for arbitrarily short waves if I considered the electron movements and electron collisions in a metal without the simplifying assumptions applicable only to long wavelengths. This should also lead to an understanding of the relationship between the constant $\lambda_m T$ and certain quantities related to free electrons. For example, the following relationship might exist. The maximum wavelength λ_m might have the value $\lambda_m = a \cdot c^2 / v^2 \cdot R$, where R is the radius of an electron, v the mean velocity of the electron, c the velocity of light, and a is a numerical constant.

I have struggled almost continuously with this problem in recent years until I finally came to the realization that it would be impossible to reach my goal in this way.[15]

Just as Planck did as late as 1906 (see page 34), Lorentz also originally expected that the equipartition theorem of statistical mechanics would prove inapplicable to systems containing "aether" in addition to ponderable matter: "Thus, it should be possible to show that the Jeans theory does not have general validity. I had harbored this hope for a long time."[15]

But Lorentz succeeded in deriving the equipartition theorem for arbitrary systems from the general fundamental equations of mechanics and electrodynamics. This approach also inevitably led him back to Jeans' radiation formula. Nothing was left for Lorentz but to conclude that Jeans, "strange as it seems, may be correct when he states that our investigations did not reach equilibrium and that the observed radiation had not been 'black'."

3. The Rome Lecture of 1908

Planck, for one, looked forward with great interest to Lorentz's announced lecture on radiation problems at the IV. International Congress of Mathematicians in Rome. Shortly before Lorentz's departure Planck wrote him on April 1, 1908:

Of course, I shall be extremely interested in learning from your lecture in Rome what your thoughts are concerning the urgent question of the distribution of energy between aether and matter. It seems quite plausible to me that, unless new hypotheses are introduced, the electron theory will necessarily lead to Jeans' conclusions, and I believe that it will be very useful if this point is expressed emphatically. In this respect there is no essential difference of opinion between us. But the controversial question is this: Does a true *equilibrium* exist for a finite energy distribution between matter and aether? The Jeans theory says no. But I believe that experience says yes. And if this is really so, then the electron theory must obviously be supplemented by a new hypothesis in order to prevent that, in the course of time, all energy transfers from matter into aether.[19]

This letter of Planck was too late to change the tenor of the lecture in Rome of April 8, 1908, entitled "Le partage de l'energie entre la matière pondérable et l'éther." Comparing the theories of Planck and Jeans, Lorentz stated that both had their advantages and their weaknesses:

The theory of Planck is the only one that would provide us with a formula in accord with experimental results; but we could accept it only with the stipulation that we completely rework our basic conception of electromagnetic phenomena. This can be seen if we consider how a single electron in an arbitrary state of motion emits radiation of all wavelengths. It is apparent that in this case it is impossible to apply the hypothesis of energy elements whose magnitude depends on frequency. On the other hand, Jeans forces us to ascribe the agreement between observations and the laws of Boltzmann and Wien to an unexplainable fortuitous coincidence.[20]

It was to be expected that Lorentz's nature would tend in this dilemma to give more weight to the logical arguments, which definitely seemed to favor Jeans' theory. But then he had to face up to the difficulty of disagreement between theory and experimental results:

I must still address myself to the question how the Jeans theory, which involves no constants other than the single coefficient k, can take into account the peak of the radiation curve which has been demonstrated by experiments. The explanation given by Jeans—which is really the only one that can be given—is that the maximum is illusory; its existence is simply an indication that it has not been possible to realize a body that is black to short wavelengths.[21]

In an effort to clarify this possibility, Lorentz chose the example of a linearly oscillating electron whose acceleration is given by

$$\mathrm{d}^2 x/\mathrm{d}t^2 = a^2 b^2 /(a^2 + b^2 t^2)^{3/2}.$$

In order to arrive at the emitted electromagnetic radiation, Lorentz developed the expression into a Fourier integral. According to him, the amplitudes of individual oscillations are then approximated for high frequencies ν by

$$\frac{1}{ab^2} \sqrt{\frac{\pi b \nu}{2a}} \, \mathrm{e}^{-a\nu/b}$$

This means, however, that the amplitudes of the emitted frequencies become infinitely small for high frequencies. Thus, such a body could emit or absorb essentially no high frequencies and would not constitute a black body in this region:

It may be assumed that a similar result would be obtained by an electron moving under the influence of a force that obeys a different law and that the absorption as well as the emission would thereby become extremely small. Thus, it may be quite likely that the body which Lummer and Pringsheim used and which corresponded completely to a black body for long wavelengths would have a far lower emissivity for short wavelengths than such a black body.[22]

4. Adverse Criticism

While Lorentz did not completely reject Planck's theory in his Rome lecture—he was too careful for that—he did make it quite clear that he expected a solution to the radiation problem through a revision of the experiments. His presentation closed with these characteristic words: "Fortunately, we can hope that new experimental determinations of the radiation function will permit a choice between the two theories."[20]

Since by 1908, Planck's radiation formula had been verified experimentally hundreds of times in all spectral and temperature regions, such statements were bound to raise violent criticism. In particular Willy Wien* raised strong objections to Lorentz's views. He turned to Otto Lummer, Ernst Pringsheim (see p. 40), Arnold Sommerfeld, and even to Lorentz himself and probably also to Planck. Thus, Wien wrote to Sommerfeld on May 18, 1908:

I was extremely disappointed by the lecture which Lorentz presented in Rome. That he came up with nothing more than the old theory of Jeans without adding any new point of view, I thought was rather skimpy. Besides, there is the question whether the Jeans theory is open to discussion in the experimental field. In my opinion, it is not worthy of discussion because actual observations show enormous deviations from Jeans' formula in a region where it is easy to check to what extent the radiation source differs from a black body. What purpose is served by submitting these questions to mathematicians† since they can provide

---

*In Planck's judgment, Wien had an excellent grasp, matched by few physicists, of both the experimental and the theoretical aspects of scientific knowledge. This applied particularly to radiation problems. Again and again, Wien contributed far-reaching comments in the debates centering around the quantum concept.

†It is a fact that mathematicians played practically no role in the development of the quantum concept, in characteristic contrast to their role in the problems that arose almost at the same time in connection with the special theory of relativity. The development of the latter was greatly furthered by the interest of Henri Poincaré, Hermann Minkowski, and Felix Klein. A mathematically complete description of the quantum theory was not developed until quite late (1925/28), and its development was based pri-

no judgment in this matter? It further strikes me as odd to show any preference for Jeans' formula which agrees with no experimental results, just because it permits retention of infinitely variable electron vibrations. And what about the spectral lines? In this instance Lorentz has not shown himself to be a leader of science.[23]

Lummer and Pringsheim, to whom Willy Wien pointed out Lorentz's exposition, raised massive objections in the *Physikalische Zeitschrift*: "If we examine the Jeans-Lorentz formula, we see at first glance that it leads to completely impossible consequences which are in cross conflict not only with the results of all observations of radiation but also with our everyday experience. We might therefore dismiss this formula without further examination were it not for the eminence and authority of the two theoretical physicists who defend it."[24]

Lummer and Pringsheim now gave a simple example which strikingly illuminates the impossibility of the Rayleigh-Jeans-Lorentz radiation law with its linear relationship between spectral intensity and temperature: Melting steel ($T \approx 1700°$ K) emits light of "blinding intensity"; a black body at room temperature ($T \approx 300°$ K) should then emit at least one sixth of this light, but this is obviously not the case.

Under the impact of such massive criticism, Lorentz recognized "the error" and "mistaken thoughts" in his address. He revised his views in favor of Planck's radiation theory and, apparently quite independently of Lummer and Pringsheim, carried out a similar simple demonstration. On June 6, 1908, at about the time when Lummer and Pringsheim were submitting their note to the *Physikalische Zeitschrift* (published July 15, 1908) Lorentz wrote to Willy Wien:

---

marily on the analysis of physical phenomena. It is the opinion of the present author that here is a significant reason why theoreticians like Lorentz (as well as Planck), with a rather one-sided orientation toward mathematical logic, could not easily find ways to treat these problems. On the other hand, a purely experimental physicist like Ernest Rutherford did not possess the prerequisites for important contributions (aside from the purely experimental) to the development of the quantum concept.

By the way, the matter became clear to me as a result of the following simple reflection. According to Jeans' theory, the radiation of a black body would be directly proportional to temperature for a given wavelength; thus, it should be [about] five times smaller at 15°C than at 1200°K. A polished silver sheet which might reflect about 93% of incident yellow light and thus absorb 7% should therefore show an emissivity for yellow light equivalent to 1.4% of the emissivity of a white-hot body. The silver sheet should therefore be visible in darkness at 15°C. Thus, we should really dismiss the Jeans theory as a valid explanation for the observed phenomena, and we are left with only the theory of Planck. Do not think that I do not respect it, quite the contrary, I admire it greatly for its boldness and success.[15]

5. The Dialogue between Planck and Lorentz

On November 21, 1908, Max Planck could joyously state in a letter to Lorentz, "that our views are now more similar" and on July 10, 1909 Planck reiterated that "we appear to agree that the energy elements $h\nu$ play a certain role in the laws of thermal radiation."[25]

They also agreed that the ideas of Einstein (and of course those of Stark as well) concerning "physically localized quanta" should be rejected as going too far. Planck wrote:

The fact that the light quanta *cannot possibly* maintain their individuality during propagation in free aether was demonstrated so convincingly in the second part of your letter that I believe that here there also can be no doubt. We would have to reject all of optics, particularly the theory of interference, refraction, and diffraction if we were to admit a separate existence for light quanta in free aether.[26]

In the same letter, dated June 16, 1909, Planck also discusses Lorentz's views as the latter had explained them in his letter of April 9/11, 1909; at the same time, he takes a critical stand. The following quotation follows immediately after the letter portion reproduced above:

Nonetheless, you tend toward the view that h is a constant of aether. I also can see the advantage of this viewpoint which in one stroke appears to eliminate the difficulties of the energy exchange between free electrons and aether; but in their place I see new difficulties which to me seem unsurmountable. If h is really a constant of aether, would it still be possible for Maxwell's equations to maintain their validity in free aether? You have yourself stressed the hopelessness of any steps which would challenge these fundamental equations. Now you suggest a way out which maintains Maxwell's equations by simply limiting the degrees of freedom of aether, rigidly coupling certain groups of adjacent vibrations. In principle, this of course leads to the desired result. But what bothers me is the thought that the type of coupling, as you yourself stress, would then depend on the boundaries of space. What would become of the "free" aether? And how would aether behave in unlimited space? I, for one, cannot reconcile such a direct influence of the (arbitrarily far removed) bounding surfaces on the processes in free aether, with the basic thought behind Maxwell's theory which excludes any type of action at a distance.[26]

In a subsequent letter dated July 10th, Planck again addressed himself to Lorentz's views (with a quotation from his letter of April 9):

You tend to bring the significance of h into relation with a limitation of the degrees of freedom of aether in such a way that each degree of freedom "absolutely refuses to accept energy in any way except in $h\nu$ portions."[25]

Thus, having failed to make progress with his electron theory around 1909, Lorentz, like Jeans and Rayleigh, became inclined to view the vibrations of aether separately (i.e., not interacting with oscillators or electrons). In this way, young Peter Debye was able to achieve an "exceedingly simple" derivation as well as a greater understanding of the Planck radiation formula, leaning more toward the views of Einstein than those of Lorentz. In characteristic contrast to Lorentz, he did not attempt to conceive some mechanistic model for aether vibrations but from the outset abstained from any "electromagnetic or mechanical explanation of h," to use the words of Sommerfeld's proposals of 1911.

For a long time, Planck simply could not comprehend such a point of view. In his long letter to Lorentz dated July 10, 1909, Planck explained his position in detail:

One more word about the case of a cavity devoid of matter or electrons and filled only with radiation. In my view, such a situation would offer *no possibility* to even understand the approach of a steady-state condition, to a normal energy distribution. For this, something like a carbon particle is absolutely essential. You assign only a very subordinate role to this but in my opinion it is the very essence of the matter. How would it be possible to determine whether or not the spectral energy distribution in cavity radiation is normal if one were not to use a carbon particle to determine whether or not it has an influence on the energy distribution? . . . To me it seems impossible to even *define* temperature, entropy, or probability for pure cavity radiation without considering the effect of this radiation on emitting and absorbing particles.[25]

In Planck's view, the establishment of a normal energy distribution in a cavity requires the Planck oscillators which exchange energy with radiation; in addition, free electrons are necessary to effect the energy interchange between the resonators. Yet Planck left open the question of how this was to take place. In 1909, he considered the possibility that in collisions of electrons with the oscillators, energy is simultaneously transferred to aether. The following assumption, Planck wrote, appears to agree with the facts:

The energy exchange between electrons and free aether always occurs only in integer multiples of the quantum $h\nu$. This is true for free electrons as well as for electrons that vibrate around equilibrium positions, such as my resonators. For slow changes in the velocity of free electrons, ν is so small that $h\nu$ essentially vanishes. Thus, we have Hamilton's ordinary differential equations. But in the course of rapid velocity changes, such as occur during collision (generation of thermal radiation, x rays, γ rays), the light quantum comes into play. As soon as the energy is transferred to free aether, Maxwell's equations are completely valid.[26]

In the same letter, Planck continues that "since the exchanged energy is equal to the difference between the emitted and absorbed energy, the absorbed energy can remain entirely constant. Only the emitted energy would have to change in steps." This remark shows that Planck was considering the hypothesis of quantum emission, which he published at the beginning of 1911, as early as the middle of 1909.

Arnold Sommerfeld took up Planck's initial ideas and developed them into his quantum-theoretical "fundamental hypothesis" (see p. 113). According to him, the interaction between electrons and atoms is definitely and uniquely controlled by Planck's quantum of action, and in his opinion this process should also explain black-body radiation.

While Lorentz in 1908 still used Hamilton's ordinary differential equations as a theoretical basis Planck, on the other hand, believed in November 1908 "that Hamilton's equations simply do not apply to a collision between an electron and an oscillator since here the quantum of action h plays a role."

Planck felt that the interaction between electrons and atoms invalidates the principle of continuity of natural processes. While legends persist that Planck already clearly understood the consequences of his law of black-body radiation by the turn of the century, it is definitely true that the matter did become clear to Planck in 1908. In 1910, when Planck again began to publish on matters related to the radiation problem after a long period of silence, he wrote with laudable clarity:

The constant h cannot be reconciled with the Jeans theory. Rather, my formula reduces to Jeans' if h is taken to be infinitely small. This circumstance appears to be a definite indication that certain elementary radiation processes which are considered continuous by the Jeans theory are actually discontinuous, and therefore correction of the theory must be approached from this direction. In my opinion, it is thus not yet necessary to give up the principle of least action which has justified its universal application so often, but we cannot maintain the general validity of Hamilton's differential equations. These were derived from the principle of least action under the assumption that all physical processes derive from changes that are continuous in time. If radiation processes are no longer made to obey Hamilton's differential equations, the ground is removed from under Jeans' theory, which stands and falls with these differential equations, and the theory's radiation formula is stripped of its universal significance, thus eliminating the conflict between theory and practice.[27]

6. Lorentz–A Leader in Science?

"Despite the active participation and penetrating understanding with which he took notice of every significant innovation," stated Planck in his commemorative address on Lorentz, "he always maintained an extremely careful and reserved attitude towards such innovations, quite appropriate for a true classical physicist . . . ."[28] Planck's judgment applies to Lorentz's attitude both toward the special theory of relativity and toward quantum theory. In his letter of May 18, 1908, to Sommerfeld, Willy Wien, still under the influence of Lorentz's lecture at the Congress of Mathematicians in Rome, denied Lorentz "in this instance" the role of "leader of science." From our perspective of history, must we still credit Lorentz with such a role in the development of the quantum theory?

Lorentz did not make any significant new contributions; in this respect he was left far behind Planck, Einstein, Sommerfeld, Nernst, and Bohr and perhaps even behind Stark and Haas, both of whom can hardly be placed in the same category as Lorentz with respect to their scientific importance. But Lorentz time and again produced unifying scientific surveys which thoroughly analyzed the state of development in a given field. This he did for the quantum problem in his Wolfskehl Lectures in Göttingen, 1910, and at the first Solvay Congress in Brussels, 1911.

In these widely respected reviews, Lorentz considered all those contributions that seemed significant to him, and his evaluation was used as a criterion: Whatever was accepted by Lorentz was accepted; and whatever was rejected by him was rejected or at best labeled "controversial." In these thorough and conscientious analyses, Lorentz demonstrated a masterly grasp of his subject.

Lorentz always took a serious view of the obligation imposed on him by the authority vested in him, earning his position again and again. His correspondence with various professional colleagues, now available to historians, clearly shows his role as a true *praeceptor physicae* in providing clarifications to many problems and sacrificing his time to give detailed opinions.

Planck played a role quite similar to that of Lorentz. With respect to the quantum problem there was the significant difference that Planck himself had been responsible for the initial developments because of his series of outstanding original contributions. In later years, the major significance of both Planck and of Lorentz lay in their delivery of survey reviews which reflected the status of research, familiarized younger physicists with the importance of the problem and pointed out some definite direction for their research.

In German physics, Planck was often called in as the umpire. At international conferences, this function devolved upon Lorentz who, as a Dutchman, acted as coordinator between his colleagues from Germany, France, and England. Thus he was appointed chairman of the first Solvay Congress in Brussels. In later years it was hardly possible to envision any other personality at the head of an international convention. "Lorentz was the perfect honorary chairman" according to Planck; "He led with the simple dignity characteristic of him, at the same time controlling debates with intelligent circumspection."[29]

Despite Lorentz's great contributions even to quantum theory, his extreme initial caution, as expressed particularly in his review lecture in Rome on April 8, 1908 greatly inhibited its development in the critical early years. Since the physicists of that time informed themselves primarily through publications in their own language and recognized few authorities outside of this sphere, the quantum concept remained for many years an internal German matter as a result of its rejection by Jeans and Rayleigh and the skeptical attitude of Lorentz, thereby denying it a foothold in England and France. Only in 1911 did the quantum theory cross the borders, but by then this happened with considerable force (see page 142). By "border" we mean the language barrier; it is in this sense that we are calling Einstein a "German" physicist.

The careful and reserved attitude of Lorentz corresponded to his personality; but it was undoubtedly also typical of the general trend of thought near the turn of the century; both state and science were regarded as edifices which, after a hectic building phase, were now considered permanently and solidly established.* Whoever shook the foundations, regardless of his motives, identified himself as an outsider.

Rayleigh, Jeans, Lorentz, and even Planck hardly differed from each other in this conservative view. In the first years, the difference between Planck and the others was only that the former had been familiar with the natural physical constant h since May 1899, and he knew that its numerical value could be determined with great precision from radiation measurements. As far as he was concerned, h was a given fact and all that was left to do was to explain its existence, uncomfortable though this might be.

After his lecture in Rome, Lorentz was won over to Planck's conviction. Yet despite this, both maintained their basically conservative positions. Like Lorentz, Planck also wanted to "proceed as circumspectly as possible in the introduction of the quantum of action h into the theory."[30] By approaching the problem with such attitudes, Planck as well as Lorentz barred their own way to new insights which, as we now know, were in conflict with a number of preconceptions of the nineteenth century.

This explains why the detailed discussions between Planck and Lorentz, while of interest to the historian, did not produce any results. It was left to Albert Einstein to show that Planck's quantum of action plays a fundamental role not only in the radiation formula but also in numerous other physical phenomena. Einstein cut through the restraints of the radiation problem and opened the way to an analysis of the quantum problem.

References

1. Friedrich Paschen, letter to Heinrich Kayser, Staatsbibliothek Preussischer Kulturbesitz. Sign. Darmst. F 1 e 1893, December 30, 1900.

*"Heavier artillery than that will be required to unsettle the still firmly established edifice of the electromagnetic theory of light" wrote Planck in 1910.

2. Ibid., January 3, 1901.

3. Heinrich Kayser, Handbuch der Spektroskopie, vol. 2, Leipzig 1902, p. 122.

4. Eugen Lommel, Lehrbuch der Experimentalphysik, 12th and 13th editions, Leipzig 1906, p. 540.

5. Orest Danilovitch Chwolson, Lehrbuch der Physik, vol. 2, Braunschweig 1905 (translated from the Russian), p. 230.

6. Eduard Riecke, Lehrbuch der Physik, vol. 2, 4th edition, Leipzig 1908, p. 714.

7. Leo Graetz, "Wärmestrahlung," in: Handbuch der Physik, vol. 3, Leipzig 1906, p. 388.

8. Lord Rayleigh, Scientific Papers, Cambridge 1903, vol. 4, p. 484.

9. Lord Rayleigh, Nature, vol. 72, 1905, p. 250.

10. James Jeans, Philosophical Magazine, Ser. 5, vol. 10, 1905, p. 98.

11. Ref. 8, vol. 4, p. 485.

12. James Jeans, Philosophical Transactions of the Royal Society, Ser. A, vol. 196, 1901, p. 397.

13. Ref. 10, p. 97.

14. Max Planck, Vorlesungen über die Theorie der Wärmestrahlung, Leipzig 1906, p. 204.

15. Hendrik Antoon Lorentz, letter to Willy Wien, Deutsches Museum München, Autograph Collection of the Library, June 6, 1908.

16. Ernst Hagen and Heinrich Rubens, Annalen der Physik, vol. 11, 1903, pp. 873-901.

17. Max Planck, Physikalische Abhandlungen und Vorträge, vol. 2, Braunschweig 1958, p. 57.

18. Hendrik Antoon Lorentz, Collected Papers, vol. 3, The Hague 1936, p. 168.

19. Max Planck, letter to Hendrik Antoon Lorentz, Algemeen Rijksarchief, The Hague, April 1, 1908.

20. Ref. 18, vol. 3, 1934, p. 341.

21. Ibid., p. 339.

22. Ibid., p. 340.

23. Willy Wien, letter to Arnold Sommerfeld, Deutsches Museum München, Autograph Collection of the Library, May 18, 1908.

24. Otto Lummer and Ernst Pringsheim, Physikalische Zeitschrift, vol. 9, 1908, p. 449.

25. Ref. 19, letter, July 19, 1909.

26. Ibid., letter, June 16, 1909.

27. Ref. 17, p. 239.
28. Ibid., vol. 3, p. 346.
29. Ibid., p. 347.
30. Ibid., p. 247.

3

Albert Einstein   Light Quanta and New
                                       Quantum Phenomena
                                       (1905–1910)

1. The "Technical Expert, Third-Class"

In 1905, Albert Einstein, then 26 years old and employed as
"technical expert, third-class" at the Swiss patent office in Bern,
published three articles in Volume 17 of the German Annals of
Physics (Annalen der Physik). Any one of these articles would
have assured the author's lasting fame. Their tremendous influence
extends far beyond the natural sciences to all of man's thoughts
and ideas.

In his theory of Brownian motion[1] he provided direct and
definitive proof for the atomic structure of matter based entirely
on classical theory. Due to thermal agitation, microscopic
particles in a liquid suspension carry out movements that can be
observed visually by microscope. The equation which Einstein
derived for the displacements of these particles was verified experimentally by Perrin. These displacements increase toward smaller particles, and extrapolation to molecular size yields the thermal
motion of a molecule, demonstrating that the invisible molecule
is just as real as the suspended particle observed under the microscope. Thus the objections raised by positivists like Ernst Mach
and Wilhelm Ostwald against the existence of molecules were
conclusively demolished.

In his paper "On the Electrodynamics of Moving Bodies,"[2]
Einstein justified his special theory of relativity by a penetrating
analysis of the concepts of space and time. This led him within a
few months to postulate a general equivalence of mass and
energy,[3] the famous equation $E = mc^2$.

The special theory of relativity sealed the fate of the earlier
principles of the conservation of mass and of the conservation of
energy. These were now replaced by a generalized formula for the
conservation of energy to which the rest mass energy must be
added.

As Einstein stated as early as 1906, a change in mass due to
conversion of energy cannot occur in chemical but only in

radioactive reactions. In 1907 he proposed a calculation of the energy released, using the difference between the atomic weight M of a radioactively decaying atom and the sum of the atomic weights of the end products Σm. By a comparison with measured values it should then be possible to verify the validity of the mass-energy relationship. "Whether the method can be successfully applied depends primarily on the existence of radioactive reactions for which $(M - \Sigma m)/M$ is not too small with respect to unity."[4] Thus already at that time attention was directed toward reactions—as yet only for purposes of measurement—for which the relative mass defect is as large as possible. Today this physical value plays an important role in nuclear physics as a measure of stability for the atomic nucleus and is equally significant in such applications as the atomic reactor and the atomic bomb.

Einstein's third article in the famous Volume 17 of the *Annalen der Physik,* "On a Heuristic Viewpoint Concerning the Production and Transformation of Light,"[5] brought a completely new viewpoint to bear on discussions of the radiation problem. He provided the essential new stimuli which conservative thinkers like Lord Rayleigh, James Jeans, and Hendrik Antoon Lorentz were unable to produce. At last, the "radiation problem" had become the "quantum problem"; in other words, most physicists began to realize that Planck's quantum of action had wider significance than in connection with the radiation equation.

A historian of physics is tempted to trace in detail the process which slowly shaped Einstein's insights, to follow his turns and blind alleys along the way. But Einstein, at least in his early years, possessed such extraordinary creative powers, enabling him to reshape his tentative conceptions so quickly that it was impossible for the present author to do justice to the evolution of Einstein's ideas in view of the meager facts at hand. This situation might change when Einstein's many earlier letters to Johann Jacob Laub are made available to historical research. Today we must still be content to accept Einstein's fully developed conceptions as they appeared between 1905 and 1909.

Einstein, whose independent thinking and intellectual irreverence later became almost legendary, took an extremely critical view of the "hallowed edifice of scientific knowledge." It seemed incredible to him what others were willing to accept as fact:

A fundamental difference of form exists between the theoretical views which physicists have embraced concerning gases and other ponderable bodies and Maxwell's theory of electromagnetic effects in a so-called vacuum. Thus, while we consider the state of a body as completely determined by the positions and velocities of a very large but finite number of atoms and electrons, we use continuous three-dimensional functions to determine the electromagnetic state existing within some region, so that a finite number of dimensions is not sufficient to determine the electromagnetic state of a region completely. According to Maxwell's theory, energy is viewed as a continuous three-dimensional function for all purely electromagnetic phenomena, including light, while according to the present view of physicists the energy of a ponderable body must be represented by a summation over the atoms and electrons. The energy of a ponderable body cannot be subdivided into an arbitrarily large number of arbitrarily small parts while the energy of a light beam radiated from a point source is continuously distributed over a steadily increasing volume according to Maxwell's theory of light (or, more generally, according to any undulatory theory).

The undulatory theory of light, which operates with continuous three-dimensional functions, applies extremely well to the explanation of purely optical phenomena and will probably never be replaced by any other theory. However, it should be kept in mind that optical observations refer to values averaged over time and not to instantaneous values. Despite the complete experimental verification of the theory of diffraction, reflection, refraction, dispersion, and so on, it is conceivable that a theory of light operating with continuous three-dimensional functions will lead to conflicts with experience if it is applied to the phenomena of light generation and conversion.[6]

Einstein appears to have set himself the task of eliminating that "difference of form" with respect to the atomic structure of matter by introducing a corpuscular structure of electromagnetic radiation. He thereby expected to avoid the difficulties which Planck encountered in the theory of black-body radiation.

This view is clearly demonstrated in a manuscript written about

1910. Maxwellian electrodynamics had "served science well," according to Einstein, because it provides correct time-averaged values, as shown by optical observations and by the Stefan-Boltzmann and Wien laws, which are based on Maxwell. Nevertheless, according to Einstein, Maxwell's electrodynamics structured on Faraday's field concept is only an intermediate (and perhaps unnecessary) stage in the development, because continuous distribution of physical quantities (above all energy) is a fiction that cannot be maintained in the light of more refined observations that do not depend on average values alone. Einstein deliberately related his ideas to pre-Maxwellian and pre-Faraday electrodynamics, which did not recognize the field concept. The pertinent section of Einstein's manuscript in response to Planck's view is as follows:

Furthermore, you [Planck] see a weakness of the quantum concept in that it does not show how static and stationary fields should be viewed. In this connection I am decidedly of the opinion that the development of relativistic electrodynamics will lead to a different localization of energy than that which we presently assume without good reason. To me it seems absurd to have energy continuously distributed in space without assuming an aether. Also, it can be easily shown that the localization of energy inherent in the old theory of action at a distance is reconcilable with Maxwell's theory. I intend to publish this shortly in connection with other matters. While Faraday's representation was useful in the development of electrodynamics, it does not follow in my opinion that this view must be maintained in all its details.[7]

2. The Light-Quantum Hypothesis

Einstein observed that the two critical formulas on which Max Planck based his theory of black-body radiation contradict each other or at least cannot be simply considered reconcilable with each other.

Planck's equation

$$u = \frac{8\pi\nu^2}{c^3} U \qquad\qquad (1)$$

was derived from the laws of classical Maxwell-Lorentz electrodynamics. In this derivation, it was assumed without question that amplitude and energy of a Hertzian oscillator are continuously variable.

According to statistical thermodynamics, the equipartition theorem applies to the average energy of a linear oscillator

$$U = kT \tag{2}$$

while Planck derived the expression

$$U = \frac{h\nu}{e^{h\nu/kT} - 1}, \tag{3}$$

assuming discrete energy steps.

The dilemma consisted in the simultaneous application of equations (1) and (3) which were derived from contradictory assumptions. Here James Jeans, Lord Rayleigh, and, until 1908, H. A. Lorentz took a consistently classical stand. They rejected out of hand the quantum-theoretical equation (3) and thus also Planck's radiation formula, at least as far as the physical end state or equilibrium state is concerned. In their opinion, the measured radiation distribution was to be understood as an intermediate state which had not been closely examined.

This was a way out which Max Planck specifically rejected. Instead, it was his view (and at times also that of Lorentz) "that the . . . difficulty is only the result of an unjustified application of the expression for the uniformity of energy distribution to all independent variables of state."[8] In Planck's opinion proper application of classical statistical mechanics would lead to (3) rather than to (2).

Einstein rejected the views of Rayleigh and Jeans as well as that of Planck. He emphasized "that this equation [2] is an unavoidable consequence of statistical thermodynamics"; despite this, equation (3) is correct. It is that that constitutes the break with classical physics. Einstein's unconventional thinking did not constrain him, as did Rayleigh's, Lorentz's, and Planck's views, to proceed as "conservatively as possible in the introduction of the quantum of action h into the theory."[9] From the beginning, Einstein viewed Maxwell's equations as valid only for time-averaged values. Furthermore there exist natural laws, like (3), which stand outside Maxwell's theory, but also conclusions that stem from the theory and that are in agreement with reality. Among these are the interference laws, the Stefan-Boltzmann law, and the Wien displacement law; according to Einstein, formula (1) should be added to these: "Although Maxwell's theory is *not* applicable to elementary oscillators, the average energy of an elementary oscillator located in a radiation cavity is equal to that calculated by means of Maxwell's theory of electricity."[10]

Einstein certainly recognized that this statement was by no means self-evident. It "would be immediately plausible if . . . [the energy element $\epsilon = h \cdot \nu$] were small with respect to the average energy U of an oscillator in all portions of the spectrum; but this is by no means true."[10] For within the range of validity of Wien's radiation formula, Einstein writes $U/\epsilon = \exp(-h\nu/kT)$.[11] This quantity should (if the above statement were true) be very much greater than unity but it is, on the contrary, much smaller.

Einstein now considered the radiation entropy within the region of validity of Wien's radiation formula. He introduced the quantity η (radiant energy in volume v, i.e., $\eta = u \cdot v$) and obtained

$$s = -\frac{k\eta}{h\nu}\left[\ln\frac{\eta c^3}{v8\pi h\nu^3\,d\nu} - 1\right].$$

For a fixed energy η the radiation now expands throughout the volume v_0. This leads to the following expression for the relationship between radiation entropy and the volume:

$$s - s_0 = k \cdot \ln\left[\left(\frac{v}{v_0}\right)^{\eta/h\nu}\right].$$

Since, on the other hand, the probability that n molecules of an ideal gas are not located within the total volume v_0 but rather in the partial volume v is given by $W = (v/v_0)^n$, comparison with the Planck-Boltzmann formula $s - s_0 = k \cdot \ln W$ immediately shows that $\eta/h\nu$ signifies the number of "energy quanta." Such an expression for entropy or probability is equivalent to the validity of the general gas equation $pV = nRT$ for the "particle" in question, as was pointed out by Einstein.

Einstein was therefore able to conclude that "monochromatic radiation of low density (within the region of validity of Wien's distribution law), behaves with respect to thermal phenomena as if it were composed of independent energy quanta of magnitude $(R/N_0) \beta\nu$."[12]

3. Further Evidence Supporting the Hypothesis

Einstein found little support for the light-quantum hypothesis which he proposed in 1905.* Even followers of the quantum concept generally rejected it. Thus, Lorentz wrote to Wien on April 12, 1909: "While I no longer doubt that the correct radiation formula can only be reached by way of Planck's hypothesis of energy elements, I consider it highly unlikely that these energy elements should also be considered as light quanta which maintain their identity during propagation." It is well known that Planck still rejected the light-quantum hypothesis as late as 1913.

On July 6, 1907, Planck wrote to Einstein: "I look for the significance of the elementary quantum of action (light quantum) not in vacuo but rather at points of absorption and emission and assume that processes in vacuo are *accurately* described by Maxwell's equations. At least I do not yet find a compelling reason for giving up this assumption which, for the time being, seems to be the simplest."[13]

In the same letter Planck mentioned that he would spend his vacation the following year in Switzerland and "might then have

*In 1910, Max Planck mentioned only Albert Einstein, Johannes Stark, Joseph Larmor, and Joseph John Thomson as adherents of the light-quantum hypothesis.

the pleasure" of meeting Einstein personally. While Planck did spend three weeks in 1908 with his family in Axalp, he had little "time and inclination for scientific discussions."[14] The exchange of ideas between Planck and Einstein continued by way of letters; the latter described the correspondence which he had with Planck (1908 and 1909) as "broad" and "lively."

In 1908, Einstein wrote to his friend Johann Jacob Laub: "Planck is also quite congenial as a correspondent. Only he has the weakness of being unable to find his way into trains of thought that are strange to him. This explains why he raises the wrong objections to my latest radiation study. Yet he gave no response to my criticism. Thus, I have the hope that he has read and approved it. This quantum question is so enormously important and difficult that everyone should be working on it."[15]

Einstein himself, for one, made tremendous efforts. In order to disarm the critics of the light-quantum hypothesis (whose leader was Planck) and thus make progress in the quantum question, Einstein toward the end of 1908 studied the energy fluctuations in a radiation cavity. He used as a model his earlier considerations on the theory of Brownian motion.

The essential idea was to apply Planck's formula $s = k \cdot \ln W$ also to systems not in equilibrium, to invert the formula to $W = \exp(s/k)$ "and to make full use of the thermodynamic properties of s."[16] (The symbol s is used because we are dealing with the entropy of the radiation and not that of the oscillators.)

Einstein considers a small partial volume v; fluctuations will occur since the energy and the entropy do not maintain their mean values but show a statistical deviation up and down. The radiation entropy σ of the partical volume v is viewed as a function of radiation energy η and is expanded in powers of $\Delta\eta = \overline{\eta} - \eta$. The sum of all partial entropies has no linear component, since $\Sigma\Delta\eta = 0$. Thus,

$$s = \text{const} - \frac{1}{2} \cdot \frac{\mathrm{d}^2\sigma}{\mathrm{d}\overline{\eta}^2} \sum (\Delta\eta)^2 .$$

If this expression is substituted into $W = \exp(s/k)$, one obtains the probability distribution for $\Delta \eta$ as a Gaussian curve, with

$$\overline{(\eta - \overline{\eta})^2} = - k \Big/ \left(\frac{\mathrm{d}^2 \sigma}{\mathrm{d}\overline{\eta}^2}\right) . \tag{4}$$

Thus, Einstein found the following mean square fluctuation for the radiation energy in the partial volume v under consideration to be

$$\overline{(\eta - \overline{\eta})^2} = \frac{R}{Lk} \left[v h \overline{\eta} + \frac{c^3}{8\pi v^2 \, \mathrm{d}v} \frac{\overline{\eta}^2}{v} \right]$$

and also was aware of the expression (but never published it) for the relative fluctuation, keeping in mind that $R/k \cdot N = 1$,

$$\frac{\overline{(\eta - \overline{\eta})^2}}{\overline{\eta}^2} = \frac{v h}{\overline{\eta}} + \frac{c^3}{8\pi v^2 \, \mathrm{d}v} \frac{1}{v} . \tag{5}$$

"Thus we have achieved," wrote Einstein, "an easily interpreted expression for the mean value of the fluctuations of the radiation energy located in v."[17]

He added that the second term corresponds to the interference of wave trains known to "present theory," that is, to Maxwell-Lorentz electrodynamics. "We would have obtained only this second term if we had started from the Jeans formula."[17]

Einstein explained this term by means of a dimensional analysis; an exact derivation was given by Lorentz in 1912.[18] The second term in (5) indicates the number $Z(v)\mathrm{d}v$ of resonant oscillations within the partial volume v of the cavity.

Of particular interest is the first term which, according to Einstein, results from Wien's radiation law: "If only the first term were present, the fluctuation of radiation energy would be that which is obtained *if the radiation were composed of independently moving point quanta of energy hv.*"*

*This can best be demonstrated if η is viewed as a multiple n of hv. One can then take the fluctuations Δn and obtain, if only the first term were present, $\overline{(\Delta n)^2} = \overline{n}$ for the mean square fluctuation. This formula applies to particles that are independent of each other.

This conclusion has almost the identical wording as that used by Einstein in his first paper on light quanta in 1905. There, he had written:[12] "Monochromatic radiation of low density (within the region of validity of Wien's distribution law), behaves with respect to thermal phenomena as if it were composed of independent energy quanta of magnitude $(R/N_0)\, \beta\nu\; [=h\nu]$."

In the earlier study, Einstein had drawn his conclusions regarding light quanta from the dependence of entropy on volume, whereas now he derived them from a consideration of energy fluctuations. To this last argument, he now also added the formula for the radiation pressure fluctuations:*

If only the first term were present, the fluctuations of radiation pressure could be fully explained under the assumption that the radiation consists of mutually independent, moving, and quite compact complexes of energy $h\nu$. Here again the equation indicates that according to Planck's formula the effects of the two previously stated causes of the fluctuations behave like fluctuations (errors) having independent origins (additive linkage of the terms composing the mean square fluctuations).

In fact, the formula for the mean square fluctuation (4) contains in its numerator, disregarding constant factors, Planck's quantity $R = 1/(d^2S/dU^2)$. Thus, Einstein provided a later justification of the procedure used by Planck in his first derivation of the radiation formula on October 14, 1900. For Planck at that time had joined the two boundary cases (Wien's and Rayleigh's radiation laws) through interpolation, by adding the two expressions for R (see p. 13).

4. A Unified Theory of Light Quanta and Electrons

During the years 1908 and 1909 Einstein directed his efforts toward finding a modification of Maxwell-Lorentz electrodynamics. For although this theory does furnish, "as shown by the excellent agreement of theory and experiments in the field of optics, the correct time-averaged values which alone are directly observable, it leads necessarily to laws concerning the thermal

*But even here, Einstein mentions only the energy $h\nu$ of the light quanta; he does not yet introduce the individual impulse $h\nu/c$.

properties of radiation which are irreconcilable with experience."[19]

According to Einstein, light quanta are an expression of these "thermal properties of radiation," quanta which are of course unknown to classical electrodynamics. But, to paraphrase another of Einstein's statements, the electron is likewise "a stranger to Maxwell-Lorentz electrodynamics."

A dimensional analysis (in which Einstein followed up the work of James Jeans and Max Planck) now showed that e^2/c has the dimensions of an action, just like Planck's quantum of action h. Einstein now felt that, except for a numerical factor, $e^2/c = h$: "It seems to me that we can conclude from $h = e^2/c$ that the same modification of theory that contains the elementary quantum e as a consequence, will also contain as a consequence the quantum structure of radiation."[20] Planck had already expressed a similar supposition in a letter to Paul Ehrenfest dated July 6 1905. He had thought that from here there might be "a bridge to the existence of an energetic elementary quantum."[21] Today we consider e, h, and c as independent physical constants; we expect that a theory of elementary particles yet to be discovered will explain why e^2/c and h have the same dimensions, i.e., explain the dimensionless quantity $2\pi e^2/hc = 1/137$, the so-called Sommerfeld fine-structure constant.

Einstein called this problem "difficult" in January 1909 but nevertheless was hopeful that he could solve it: "But the number of possibilities does not appear to be so great as to cause one to recoil from this task."[22]

By July 1909 his optimism had already waned considerably. He writes in a letter to Stark: "You can hardly imagine the trouble to which I have gone in an effort to conceive a satisfactory mathematical demonstration of the quantum theory. Up to now I have been unsucessful."[23]

Even today, Einstein's goal to incorporate both light quanta and electrons into a single theory still has not been reached.* Even though Einstein's efforts were premature, it may be of interest to indicate his ideas concerning a theory of light quanta and electrons:

The fundamental equation of optics,

$$D(\varphi) = \frac{1}{c^2} \frac{\partial^2 \varphi}{\partial t^2} - \left(\frac{\partial^2 \varphi}{\partial x^2} + \frac{\partial^2 \varphi}{\partial y^2} + \frac{\partial^2 \varphi}{\partial z^2} \right) = 0,$$

will have to be replaced by an equation in which the universal constant e (probably as a squared value) will also appear as a co-efficient. The desired equation (viz., the desired system of equations) must be dimensionally homogeneous. It must revert into itself with application of the Lorentz transformation. It cannot be linear or homogeneous. It must—at least if the Rayleigh-Jeans law is really valid in the limit for small v/T—lead in the limit to the form $D(\varphi) = 0$ for large amplitudes.[22]

Einstein's attempt to find a "unified theory" for light quanta and electrons was destined to fail; his thoughts, however, concerning the connection between Planck's constant h and the electrical elementary quantum e were taken up by Willy Wien, Arthur Erich Haas, Arnold Sommerfeld, and others and were therefore historically fruitful (see p. 90).

In 1910, Einstein developed even more radical ideas which, however, were not published. They are known to us through a letter addressed to Johann Jacob Laub dated November 4, 1910: "At this time I have high hopes of solving the radiation problem without the use of light quanta. I am most curious to see how this will come out. It will not be possible to maintain the energy principle in its present form."[24] Only a few days later, Einstein was again forced to admit defeat: "Again there is no progress toward the solution of the radiation problem. It seems that the devil is playing games with me."[24]

*In Heisenberg's opinion some future theory must clear up "at a single stroke" the entire spectrum of elementary particles according to the mutual transformability of elementary particles.

5. The "Groups of Phenomena Related to the Transformation of Light"

The light-quantum concept was viewed by other physicists of that time as the most radical attempt to derive the laws of black-body radiation and was therefore received with corresponding skepticism. Even the argument that the fluctuations of radiation demonstrated the quantum character of radiation (again, within the range of application of Wien's formula) was not considered convincing.

Thus, Einstein's application of the quantum hypothesis to simple physical phenomena was more fruitful than his theoretical expositions. Planck had made use of the hypothesis at a single juncture in a very extensive series of thoughts but Einstein now demonstrated how directly the quantum law is responsible for physical phenomena: "It appears to me that the observations which have been made on 'black radiation,' photoluminescence, the generation of cathode rays by ultraviolet light, and other groups of phenomena related to the production or transformation of light can be more readily understood under the assumption that light energy is discontinuously distributed in space."[25]

The examples which Einstein gave in 1905 were Stokes's rule of photoluminescence, the photoelectric effect in metals, and the ionization of gases by ultraviolet light (photoeffect in a molecule).

Einstein derived a simple formula for the photoelectric effect, in which the maximum velocity v of the electrons released depends only on the frequency ν of the light with which the metal plate is irradiated:[26]

$$\frac{mv^2}{2} = h\nu - P. \tag{6}$$

In particular, the light intensity has no influence on the (maximum) electron velocity v; this does not agree with classical theory. According to the undulatory theory of light, commented Einstein in 1909, "it is not at all clear why cathode rays generated photoelectrically or by X rays attain such high velocities, independent of radiation intensity."[27]

The first to be impressed by this connection between phenomena and quantum hypothesis was Johannes Stark (see p. 73). Despite this physically fruitful starting point, any understanding of the phenomena interpreted from the point of view of quanta really did not go beyond the stage of a plan of attack, particularly in the years 1906-1913 (see p. 124). Most of these effects were not yet sufficiently confirmed experimentally; also, attempts at interpretation offered great possibilities for conceiving ad hoc hypotheses. But Einstein, in his theory of specific heat, was able to provide early in 1907 a new link between theory and experiment which proved of great significance for the further development of quantum theory.

6. The Quantum Theory of Specific Heat
In 1906, Max Planck for the first time used the quantum-mechanical expression for the mean energy of a linear oscillator as part of his "theory of thermal radiation":[28] *

$$U = \frac{h\nu}{e^{h\nu/kT} - 1} \tag{7}$$

From this formula and using equation (1), Planck directly derived his radiation law; thus, equation (7) depended, as far as Planck was concerned, on a rather broad and initially still debatable justification of the radiation law.

Toward the end of 1906, Einstein provided a simplified derivation. In a manner of speaking the linear oscillator emerged, as it were, from the radiation cavity in which it had been trapped until then and started on its way to an independent existence.

Einstein postulated "that the energy of the elementary structure under consideration only assumes values that lie infinitely

*It can be assumed that by the end of 1900 Planck had already formulated this equation which, after all, represents one of his preliminary steps toward his radiation law.

close to 0, ϵ, 2ϵ etc."[29] To each of these energy states, a statistical weight of exp $(-n\epsilon/kT)$ is to be given according to Boltzmann, yielding the following mean value:

$$U = \frac{\displaystyle\sum_{n=0}^{\infty} n\epsilon\, e^{-n\epsilon/kT}}{\displaystyle\sum e^{-n\epsilon/kT}} = \frac{\epsilon}{e^{\epsilon/kT} - 1},$$

which leads directly to formula (7) with the substitution $\epsilon = h\nu$.

As Einstein fully realized, equation (7) contradicts the equipartition theorem; here we encounter the break with classical physics (see p. 54). Equation (7) draws its justification from the experimentally based radiation law of Planck. Thus, Einstein concluded:

If Planck's theory of radiation strikes into the heart of the matter, then we must also expect to find contradictions in other areas of thermodynamics between the present molecular-kinetic theory and experience, which can be eliminated by proceeding along the chosen path.[30]

Although the Dulong-Petit law, which can be directly derived from the equipartition law, was shown to apply to many materials at normal temperatures, there were innumerable exceptions, particularly at low temperatures. While earlier attempts at explanations had proved unsatisfactory, equation (7) indicates a temperature dependence in a straightforward manner; to obtain the specific heat from (7) it is simply necessary to differentiate with respect to T. If, like Einstein, we also go to 3 degrees of freedom and N atoms, we obtain the following expression for molar specific heat:

$$C = 3R\, \frac{e^{h\nu/kT} \left(\dfrac{h\nu}{kT}\right)^2}{(e^{h\nu/kT} - 1)^2} \tag{8}$$

Although Einstein knew very well that the correct formula is obtained by a summation over all existing "eigen oscillations," he was unable to carry out this calculation. Nevertheless, important conclusions can already be drawn from the simple expression (8). Einstein was fully aware that the various eigen oscillations either do or do not contribute to the specific heat, depending upon the magnitude of the corresponding frequency ν.

This is a very significant step, for spectroscopic observations had shown that a gas particle can emit many lines, each of which, according to the view of classical physics, must correspond to one degree of freedom of vibration. But in the case of specific heat, these degrees of freedom did not show up—a most peculiar situation. Lord Kelvin had specifically singled out this difficulty as one of the "19th century clouds"of physics.

Einstein now showed that at a certain temperature some vibrations could be practically inactive with respect to specific heat. Other forms of vibration become inactive in the transition to lower temperatures, corresponding to what was later referred to as the "freezing of degrees of freedom."

Since Einstein was not yet in a position to provide the spectrum of the actual frequencies—this remained to be accomplished by Peter Debye and Max Born in 1912—he limited himself to the case of a single resonant oscillation. He specifically designated this procedure as a "rough approximation," and thereby indicated the way toward a more exact treatment (see p. 132).

It was extremely important for the development of quantum theory that Walther Nernst also became interested, at about the same time as Einstein, in the behavior of specific heat at low temperatures. He approached the problem from a direction that initially appeared quite different from the problems facing the quantum concept. Nernst was well under way with his measurements, using the facilities of his large Berlin Institute, when his attention was drawn to Einstein's theory toward the end of 1909 or the beginning of 1910 (see p. 129). Thus Einstein's theory of specific heat of 1910/1911 became, next to Planck's radiation formula, a second "equally strong pillar of quantum theory" in the words of Sommerfeld.

7. The Salzburg Congress of 1909

After the three well-known papers of Einstein had appeared in the year 1905, Max Planck maintained considerable skepticism with respect to the light-quantum hypothesis. Yet he showed correspondingly greater understanding toward Einstein's theory of relativity. Planck's review at the Physics Colloquium in Berlin at the beginning of the winter semester 1905/06 was probably his first reaction to Einstein's paper on the subject.

Planck treated the questions surrounding "relativity" in many lectures and publications. When Walter Kaufmann construed his measurements of the deflection of canal rays in parallel electrical and magnetic fields as a refutation of Einstein's theory, Planck cautioned against an overly hasty interpretation of the experiments. He also energetically opposed other objections to the theory. On July 6, 1907 he wrote to Einstein:

Herr Bucherer has already warned me, through a letter, of his strong opposition to my last presentation, since he finds (without giving any reason) that the principle of relativity is irreconcilable with the principle of least action. All the more reason why I was pleased to gather from your postcard that, for the time being, you do not share his views. As long as the proponents of the relativity principle form such a small minority it is doubly important that they agree among themselves.[14]

Thus, although Planck took his place in the ranks of influential early champions of the relativity theory, he was extremely skeptical toward Einstein's hypothesis of light quanta. In the letter just quoted we find this sentence, following a discussion of questions concerning black-body radiation: "More critical than this matter, which is rather far in the background at this time, is the question regarding the admissibility of your relativity principle."

It appears that the complex of problems surrounding the theory of relativity pushed the quantum question into the background as far as Planck was concerned. Between 1906 and 1909, the principal publications of Planck dealt with the theory of relativity. The first paper again dealing specifically with quantum theory appeared early in 1910.[31] Planck again devoted himself exclusively to the quantum problem after Lorentz's lecture in Rome on April 8, 1908 (see p. 41). Fortunately, he quickly succeeded in his efforts to obtain acceptance of the special theory of relativity among the leading physicists of Germany. Important milestones along the way were the approval by Sommerfeld in 1907 and Minkowski's lecture entitled "Space and Time" at the congress of physical scientists in Cologne in 1908.

The success of the theory of relativity brought its creator, Albert Einstein, the greatest respect of his professional colleagues, with the result that his other publications, particularly those concerning the quantum problem to which he addressed himself frequently, were also given wide consideration.

Until then, Albert Einstein was personally known to only a few of the younger physicists, such as Johann Jacob Laub and Rudolf Ladenburg, who had taken the trouble to journey to Bern. When Einstein participated for the first time at a congress of German physical scientists and physicians, many of his colleagues showed unusual interest in him; his presence was undoubtedly one of the principal attractions of the convention.

The 81st congress of German physical scientists and physicians, which took place in Salzburg September 19–25, 1909, left a profound impression on all participants, particularly those in the well-attended physics section. Here are the comments, written in English, of Lise Meitner:

This congress was altogether a very impressive experience. It was attended by theoretical and experimental physicists from the entire world . . . . It was really something quite out from [sic] the ordinary, a most stimulating meeting.[32]

Einstein delivered his lecture to the physics section on September 21, 1909, at the beginning of the afternoon session. From the available figures, we can conclude that more than a hundred listeners were present at this meeting, including many of the leading German-speaking physicists.

The lecture made a deep impression on his listeners, at least the younger ones. Entitled "On the Development of our Views Concerning the Nature and Composition of Radiation," Einstein's address dealt first with the principle of relativity and then with the quantum problem. Max Born recorded that "Einstein's achievement was given the stamp of approval by the assembled body of learned men."[33]

Thus, Einstein was accepted into the inner circle of leading physicists. Planck's reply to Einstein's address reflects great respect even though Planck at the same time withheld official sanction, as it were, to the bold ideas of the young Einstein regarding the light-quantum hypothesis.* Without question, Einstein's presentation and Planck's attitude must have caused quite a sensation.

Einstein was not able to overcome the resistance to his light-quantum hypothesis by direct attack. Here are the comments, written in English, of Fritz Reiche, one of the many younger participants:

I must say, I was very much impressed by the appearance of the second term in the fluctuation formula. Though it is of course a rather indistinct proof of 'photons.' I remember of course that people were opposed and tried to find another reason or tried to give the formula another form.[34]

Paul S. Epstein also did not believe that Einstein convinced too many of his listeners with his lecture:

Heilbron: 'Do you recall whether that talk of Einstein had a great effect?' *Epstein*: 'No [great effect]. You see, the chairman of the meeting was Planck, and he immediately said that it was very interesting but he did not quite agree with it. And the only man who seconded at that meeting was Johannes Stark. You see, it was too far advanced.'[35]

Though his audience was unable to follow Einstein in his discussion of the light quantum hypothesis, they recognized the thrust of his ideas. In any event, since Einstein's first appearance in 1905 as an unknown "expert third class" at the Swiss patent office, he had become a man on whom unusual respect was bestowed. It was significant that Einstein lectured not only on the

*In his reply, Planck did not even comment on Einstein's discussion of the special theory of relativity since in his opinion no further discussion was necessary.

special theory of relativity in his review at Salzburg, a matter which he "left to lesser prophets," but rather addressed himself primarily to the quantum problem. In this way, the significance of that largely unsolved complex of problems was stressed before this great forum of assembled physicists.

The recognition which Einstein had achieved primarily by developing the special theory of relativity now induced a number of his colleagues to give serious thought to the quantum problem. Today we view the theory of relativity and the quantum theory as dealing with separate fields of endeavor: while the special theory of relativity is based on the fact that the velocity of light c is finite, the (nonrelativistic) quantum theory is a consequence of the physical constant $h \neq 0$. Thus, while the two most significant physical theories of the early 20th century have no logical connection, their historical development was nevertheless closely intertwined. The success of the theory of relativity was responsible for accelerating the development of quantum theory.

After scientific opinion had moved toward acceptance of the quantum concept during 1910 (see p. 136), many of Einstein's professional colleagues, among them Arnold Sommerfeld, expected that he would make the decisive breakthrough. Here is Einstein's reply to a letter of Sommerfeld regarding this matter, dated October 29, 1912:

Your friendly note has caused me even greater embarrassment. But I must assure you that I have nothing new to add in the quantum matter, nothing worthy of note . . . . My attention is now fully directed to the problems of gravitation . . . .[36]

Einstein is also responsible for a historical relationship between early quantum theory and the general theory of relativity in the sense that, beginning in early 1912, his attention was almost fully drawn away from quantum problems. Sommerfeld made the following regretful utterance to David Hilbert on November 1, 1912: "My letter to Einstein proved useless . . . Apparently he is so deeply involved in the problems of gravitation that he turns a deaf ear to all else."[37]

References

1. Albert Einstein, *Annalen der Physik,* vol. 17, 1905, pp. 549–560.

2. Ibid., pp. 891–921.

3. Ibid., vol. 18, 1906, pp. 639–641.

4. Albert Einstein, *Jahrbuch der Radioaktivität und Elektronik,* vol. 4, 1907, pp. 411–462.

5. Ref. 1, pp. 132–148. Reprint: Albert Einstein, Die Hypothese der Lichtquanten (Dokumente der Naturwissenschaft, vol. 7), Stuttgart 1967, pp. 26–42. Pages quoted relate to reprint.

6. Ibid., p. 26.

7. Albert Einstein, Manuscript (handwritten by Mileva Einstein), Einstein Archives, Princeton, N.J., 1910 (?).

8. Max Planck, Vorlesungen über die Theorie der Wärmestrahlung, Leipzig 1906, p. 178.

9. Max Planck, Physikalische Abhandlungen und Vorträge, Braunschweig 1958, vol 2, p. 247.

10. Albert Einstein, *Annalen der Physik,* vol. 20, 1906, pp. 199-206.

11. As Martin J. Klein has pointed out, "it is certainly significant that Einstein always wrote the magnitude of his light quanta as $(R/N_0)\,\beta\,v$ and did not use Planck's form $h\,v$. This is not merely a matter of notation, since Planck had laid emphasis on the importance of h as a basic natural constant, and Einstein's preference for the form $(R/N_0)\,\beta$ suggests that he had not accepted Planck's views" (*The Natural Philosopher,* vol. 2, 1963, pp. 82-83). For the convenience of the reader we use Planck's notation.

12. Ref. 5, p. 37. Translation by Martin J. Klein.

13. Max Planck, letter to Albert Einstein, Einstein Archives, Princeton, N.J., July 6, 1907.

14. Max Planck, postcard to Albert Einstein, September 8, 1908.

15. Carl Seelig, Albert Einstein: Leben und Werk eines Genies unserer Zeit, Zürich 1960, p. 147.

16. Max Born, Physik im Wandel meiner Zeit, 4th edition, Braunschweig 1966, p. 222.

17. Albert Einstein, *Physikalische Zeitschrift,* vol. 10, 1909, p. 189.

18. Hendrik Antoon Lorentz, Les Théories statistiques et thermodynamiques, Conférences faites au Collège de France en Novembre 1912, Leipzig and Berlin, 1916, p. 114. See also Martin J. Klein, The Natural Philosopher, vol. 3, 1964, p. 12.

19. Ref. 17, p. 190.

20. Ref. 17, p. 192.

21. Max Planck, letter to Paul Ehrenfest, Rijksmuseum Leiden, Ehrenfest Collection (accession 1964), July 6, 1905.

22. Ref. 17, p. 193.

23. Armin Hermann, Albert Einstein und Johannes Stark. Briefwechsel . . In: Sudhoffs Archiv, vol. 50, 1966, p. 279.

24. Ref. 15, p. 197.

25. Ref. 5, p. 27.

26. Ref. 5, p. 40.

27. Ref. 17, p. 821.

28. Ref. 8, p. 157.

29. Albert Einstein, *Annalen der Physik,* vol. 22, 1907, p. 23.

30. Ibid., p. 24.

31. Ref. 9, p. 237.

32. Lise Meitner, *Bulletin of the Atomic Scientists,* vol. 20, no. 11, 1965, p. 4.

33. Max Born, Die Relativitätstheorie Einsteins, Berlin 1920, p. 237.

34. Interviews with the Sources for the History of Quantum Physics (oral records): Fritz Reiche. Manuscripts in Niels Bohr Archive, Copenhagen.

35. Ibid., Paul S. Epstein.

36. Albert Einstein/Arnold Sommerfeld, Briefwechsel, Armin Hermann, ed., Basel 1968, p. 26.

37. Arnold Sommerfeld, letter to David Hilbert, Einstein Archives, Princeton, N.J., November 1, 1912.

4

Johannes Stark   The Search for New Quantum
                                      Phenomena
                                      (1907–1910)

1. The Motives

Johannes Stark was an extremely skillful experimentalist and had
a high degree of creative imagination; he was less inclined toward
abstract mathematical ideas which had increasingly characterized
the field of physics since the turn of the century. The first major
success of the 30-year-old physicist was his discovery late in 1905
of the Doppler effect of canal rays.

He immediately approached Lorentz and Sommerfeld for a
"theoretical explanation of the effects to be expected":[1] Stark
went to considerable efforts to work out the scientific significance
of his discovery. Thus he tried in 1906 to relate the optical
Doppler effect to Einstein's theory of relativity* and, starting a
year later, to the Planck-Einstein quantum hypothesis.

In addition to his scientific motive of obtaining a complete
theoretical understanding of the Doppler effect, Stark had personal reasons for accepting the quantum concept as early as 1907.
At that time anyone with such views was stamped as an outsider,
and Stark always felt a strong need to oppose learned opinion.
This also led to the seemingly paradoxical situation that he, who
had done so much for Bohr's theory, vehemently opposed it at
the time when all the pieces of evidence began to fit together in
1913 and rejected it as being "dogmatic."

As early as October 1907, Stark had examined physical phenomena that might be directly explained as a consequence of the
quantum hypothesis. "This hypothesis," he wrote in 1908, "is so
unusual and so strongly contradicts presently held conceptions of
the emission and absorption of light, that the skepticism with
which it is generally greeted is quite understandable. Yet I feel it

*To a first-order approximation $v/c \sim \Delta\lambda/\lambda$, the optical Doppler effect can
obviously be treated nonrelativistically; only in the second order (with its
transverse Doppler effect) does the special theory of relativity come into
play. But experimental techniques at that time were not yet sufficiently
developed to permit the measurement of such minute effects.

is wrong to show continued passive resistance and to demand that, starting from already known properties of electromagnetic oscillators, the new hypothetically introduced characteristics be explained; perhaps this demand is impossible to meet insofar as it is conceivable that the new characteristics of electromagnetic oscillators constitute a primary fundamental property just like the electrical charge and the mass of negative electrons."[2]

Stark enumerated a considerable list of physical phenomena which in his opinion confirmed the quantum hypothesis. Some of his views are correct but others involve quite untenable interpretations. Following is a list of Stark's arguments in support of the quantum concept:

I. X rays

Short-wave length limit (1907)
Concentrated energy (1909)

II. Cathode and canal rays

Photoelectric effect (1907)
Intensity minimum in the Doppler effect (1907)

III. Atomic theory

Excitation energy and emitted spectrum (1908)
Energy levels and radiation transitions (1908)
Fundamental photochemical law (1908)

2. The Three Arguments Given in 1907

In 1907, Stark cited three phenomena supporting the validity of Planck's quantum law, $\epsilon = h\nu$: the photoelectric effect, the short-wave length limit of bremsstrahlung, and the intensity minimum in the Doppler effect.* Einstein had already suggested the photoelectric effect in 1905. In his first publication on this subject,[3] Stark did not mention Einstein; however, in his subsequent paper

---

*Closely related to the interpretation of the Doppler effect is the concept of excitation energy which was later analyzed more exactly by Stark in 1908, independent of the Doppler effect.

he inserted the following footnote: "It has come to my attention that A. Einstein has given a similar interpretation to the photoelectric cathode rays."[4]

At about the same time as Willy Wien,[5] Stark discussed the short-wave length limit of bremsstrahlung resulting from the quantum law.[6] The law was verified experimentally in 1915 by William Duane and Franklin L. Hunt[7] and only then became generally known. In his *History of Physics,* Max von Laue wrote: "Because it [the quantum law] was not yet widely known when x-ray interferences were discovered in 1912, M. v. Laue had to expect more interference points from his theory than were actually observed and therefore erroneously attributed this deficit to selective properties of the crystal atoms. In fact, according to the Duane-Hunt law, the short wavelengths are missing, wavelengths which should have produced the missing points."[8]

In addition to the short-wavelength limit of the bremsstrahlung continuum and the photoelectric effect, Stark designated the Doppler effect in 1907 as the "third possibility for an experimental verification of Planck's elemental law."

In the spectral analysis of light emanating from a canal ray tube, the Doppler bands ("moving intensity," as Stark called it), displaced toward shorter wavelengths when observed against the direction of motion, can be seen in addition to the unshifted spectral line. A definite intensity minimum is located between the "intensity at rest" and the broad "moving intensity." In 1907, Friedrich Paschen determined a splitting of the Doppler band which was later verified by Stark and Steubing.

When Stark published his detailed study concerning the Doppler effect in 1906, he clearly raised the question of the origin of this intensity minimum. In October 1907, he felt that he had found a solution to the problem in Planck's quantum law: in collisions between moving canal ray particles and gas particles at rest, inelastic deformation energy can be transferred to the canal ray particles only if the energy available for this purpose is at least equal to $h\nu$, where ν is the frequency of optical vibration. Thus the kinetic energy of the canal ray particles must exceed a definite value. This explanation was accepted by Albert Einstein, Hendrik Antoon Lorentz,[9] Peter Debye, and Arnold Sommerfeld (see page 104). Einstein, for example, wrote the following postcard to Stark on December 2, 1908: "Many thanks for sending me your paper. I was particularly pleased with the application of light-quantum theory to the curve for the Doppler effect."[10]

But Stark's ideas could not be sustained. The processes involved in canal rays proved to be significantly more complex than could be supposed in 1907 and 1908. Velocity analysis showed that not all velocities occur in canal rays and that, in particular, the low velocities are absent. It was further shown that, for example in hydrogen canal rays, H_2^+ ions exist in addition to H^+ ions; this also provides an explanation for the split band.

3. Excitation Energy and the Emission Spectrum

Because of his interest in the optical Doppler effect, Stark was also concerned with the phenomenon of light emission and therefore undertook a study of the collision excitation of light-emitting canal ray particles. As early as 1907 he had already considered in detail[4] the excitation of light emission of atoms by collision with other atoms or electrons. One year later,[2] he referred to the collision of cathode-ray particles (electrons) with atoms as a new possibility for subjecting Planck's hypothesis to experimental verification. A collision generally involves elastic scattering; but according to Stark, a "resonance scattering" corresponds to any given velocity (energy) at which a fraction of the kinetic energy (which he called α or β) is transformed into electromagnetic vibrational energy.

In order to excite this electromagnetic vibration, Planck's quantum law now comes into play; according to Planck the radiated energy has a value $h\nu$. From this, Stark could immediately conclude "that in a collision between a moving atomic ion* and another particle, its characteristic electronic vibration and its corresponding series of emission lines can occur only if the kinetic energy . . . lies above a certain threshhold value which is characteristic for the atomic ion in question."[4] Thus, as early as 1907 Stark had conceived an "excitation energy" on the basis of simple (and certainly not very clear) considerations, an excitation energy which must be reached if energy is to be transferred in a collision with an atom capable of emitting radiation.

Stark now considered the complete spectrum of characteristic emission lines of an atom. If an interaction takes place between electrons of some given velocity v and atoms, then all emission lines that satisfy the relationship

$$\frac{\beta m v^2}{2} > h\nu$$

are excited. In other words, the emission spectrum ends at $\nu = \beta m v^2 / 2h$. Higher-frequency spectral lines do not exist.

In connection with this exposition, Stark described the following outline of an experiment: "Cathode rays are generated at a Wehnelt oxide cathode; their maximum velocity is determined by the effective cathode potential . . . . For different values of cathode potential, observations are made of the extent into the far ultraviolet region of the emission spectrum of the atomic ions. This extension also depends significantly on the value of the coefficient β. Let us assume the extreme case in which $\beta = 1$. The emission of a line at $\lambda = 250\,\mu\mu$ then requires a cathode ray velocity of 7.8 volts while the emission at $\lambda = 186\,\mu\mu$ requires a velocity of 10.4 volts. Personal circumstances prevent me at this time from carrying out the previously indicated investigations. Perhaps some other author will be helped in his approach to the problem by the following observations which have been completed."[11]

---

*Substitute the word atom for "atomic ion." Stark considered the radiating particle as being in motion and believed that the line spectra were emitted only by ions.

It was possible for Ernst Gehrcke and Rudolf Seeliger to show in 1912 that the emission of some definite spectral line occurs only when the electrons attain a minimum velocity.[12] This "velocity limit" proved to be quite sharply defined.

When Franck and Hertz began their electron collision experiments in 1911, they were not familiar with these studies of Stark, as can be seen from a letter written by James Franck to Stark on December 30, 1911. In it, he stated that Hertz and he only learned of the corresponding work of Stark subsequent to the publication of their paper on the "relationship between the quantum hypothesis and the ionization potential."[13]

While the ideas brought forth by Stark were still considered unusual in 1908, the intellectual climate, at least in Berlin, had changed so greatly by 1911 that Franck could accept the quantum concept as a matter of course.* The goal which Franck and Hertz had set themselves in their experiments was the development of a general "kinetic theory of electrons of gases." Initially, they chose inert gases and metallic vapors for which it was known that they cannot form negative ions but that the electrons continue to move freely.[14]

Experiments with slow electrons showed (in contrast to Townsend's theory) that noticeable electron energy losses occurred only after several thousand collisions. Thus the collisions could be viewed, at least in approximation, as elastic.† This was followed by the now famous mercury-vapor experiments: If the accelerating potential is raised to 4.9 volts, the electrons suddenly transfer all of the attained "critical" energy to the gas atoms. It was significant that those electrons whose energy exceeded the "critical" energy did not release their total energy but rather only this

---

*After 1912, Franck and Hertz based their work on Sommerfeld's "fundamental hypothesis" for the interaction between atom and electron.
†Just how necessary such preparatory investigations had been, is shown by a comparison with Stark's ideas. His still undetermined factor (within the limits 0 and 1) meant that no unique correlation between electron energy and emitted radiation was possible. Since this factor must be assumed as variable, a calibration was also impossible.

"critical" amount while retaining the rest in the form of kinetic energy.

What happens then to the energy released by the electrons to the gas atoms? The answer given by Franck and Hertz is that two processes take place. First, the mercury atom may be ionized at 4.9 ev: "The critical velocity is a characteristic that is constant for each gas and is equal to the ionization energy in the cases examined."* But they emphasized that "a large part of the collisions in which the energy $h\nu$ is transferred to the vibrating electron does not lead to ionization . . . . It could therefore be expected that collisions which do not lead to ionization but which result in an energy loss of $h\nu$ will be accompanied by emission of light of frequency ν."[15]

Franck and Hertz felt that their experiments verified this view: by bombarding mercury atoms with 4.9-ev monoenergetic electrons, monochromatic radiation is emitted at a wavelength of 2537 Å. This is the resonance line of mercury. According to Franck and Hertz, these results were proof for the quantum theory: a causal relationship had been established between absorption of the "critical value of energy" 4.9 ev by the mercury atom and the emission of the spectral line ν corresponding to this energy as given by $e\nu = h\nu$. But because of the fact that they also related this to the ionization potential, they did not succeed in providing further clarification of the atomic structure.

The only line which Franck and Hertz had observed was the mercury resonance line; no other lines in the normal mercury arc spectrum appeared. In retrospect, Gustav Hertz stated: "The fact that the sensitivity of our experiment was not sufficient to detect the inelastic collision at 4.66 ev leading to the excitation of the metastable state, was in reality a fortunate circumstance

---

*Only after the development of Bohr's atomic theory was it recognized that this 4.9 volts cannot be called an ionization potential. The actual ionization potential of mercury is around 10.3 volts and was first determined experimentally by Tate and theoretically by Bohr. Franck and Hertz, on the other hand, had measured the first excitation level.

since we would not have been able at that time to relate this energy quantum to the atomic spectrum of mercury."[16]

But this was only a fortunate circumstance in the progress of the studies of Franck and Hertz who thus were not distracted from work on other problems. In principle, here was a possibility for discovering experimentally the various atomic energy levels. The recognition that an atom is capable of assuming different energy states and undergoes a transition to a lower-energy state due to radiation is a significant step in the development that must be credited to Bohr and not to the experiment of Franck and Hertz (see p. 151).

The great untapped possibilities inherent in the method of Franck and Hertz were only later shown during the verification and the discussions of the atomic model proposed by Bohr. Bohr himself was the first to point this out in August 1915.[17] Hertz stated later: "Here is another indication of the fact that Bohr's views concerning stationary states with which we are so familiar today were strange indeed to the physicists of that time so that we [Franck and Hertz] did not even think of applying them to the mercury spectrum and to our measurements."[16]

4. Stark's Further Arguments of 1908

In 1908, Stark supplemented his previous four arguments supporting the "quantum law." He announced the "fundamental hypothesis . . . that the direct chemical action of light consists in the release of valence electrons from their binding as a result, according to the quantum law, of the absorption of more light by individual valence electrons than the amount of binding energy of these valence electrons."[18]

This is nothing else than the basic photochemical law which Max von Laue ascribes to Albert Einstein in 1912 in his *History of Physics*. In fact, when Einstein published his corresponding study[19] in 1912, Stark immediately claimed priority for himself.[20] In his reply, Einstein wrote: "I do not wish to discuss the question of priority brought up by J. Stark, particularly in view of the fact that the photochemical equivalence law is a completely self-evident deduction from the quantum hypothesis."[21]

Einstein followed with great interest the experiments in the "photochemical workshop" of Emil Warburg where, as he wrote to Warburg in 1912, "things become reality which for years I have only conceived in vague outline in my dreams."[22]

Again and again, Stark's thoughts reverted to the problem of the structure of the atom. As early as 1906 he had looked for answers in this field from possible influences of an electrical field on the spectral lines.[23] This laid the basis for his subsequent discovery of the Stark effect in 1913. He considered the "inner or ring electrons" of the atom responsible for the spectral series; he was of the opinion that these spectra were only emitted by ions. The "detachable or valence electrons" were thought by him to be the "origin of the band spectrum." In analyzing the band spectra, he developed the view in 1908 that, with respect to the atom, the electron may not only exist in a normal state (today we would say ground state) or in a state of complete detachment, but that a partial detachment may also be possible.* "In the recombination of a partially or completely detached . . . valence electron, the [corresponding potential] energy is, in our opinion, radiated as a band spectrum; the various lines of the latter correspond to the various possible phases of recombination; in their totality, they represent the band spectrum of the . . . valence electron."[24]

From this point of view, Johannes Stark wrote the following in (unpublished) notes which he prepared in 1947:[25] "The supposition that the energy difference between two positions of the atomic electron may be radiated in a single line was first expressed by me with respect to the band spectrum." More weight than to this self-evaluation should be given to a letter which a student of Sommerfeld, Paul S. Epstein, wrote to Stark in 1916:

*The fact that the band spectra were connected to the atom (and not, as would be correct, to the molecule) should not be surprising. Only in 1912 did Niels Bjerrum determine that the band spectra were spectra of molecules; a view confirmed about 1920 by the further development of Bohr's theory.

"In a lecture I delivered at the physics colloquium in Munich concerning the Bohr model immediately after its discovery, I particularly emphasized that the significant addition which Bohr made to the ideas of Rutherford and Nicholson, namely, that the emitted wavelength is determined from the available energy by the equation

$$h\nu = A_1 - A_2$$

was actually first proposed by you and was first mentioned in a study by Steubing in 1909."[26]

This judgment by Epstein should be interpreted not only as a statement of fact but also from a psychological point of view. Epstein, who had calculated the Stark effect according to the Bohr theory and had found excellent agreement with Stark's measurements, must obviously have been interested in winning Stark over to the Bohr theory. He evidently believed that this might be done by presenting the atomic models of Stark as a forerunner of Bohr's theory. Actually, with this judgment Epstein came close to the historical truth (see p. 155).

What were Stark's specific conceptions? A single electron is capable of emitting a complete band spectrum (or even several such spectra). From a state of almost complete separation, the valence electron returns to the vicinity of the atom over a very eccentric elliptical path. At the perihelion the path is strongly curved; there, electromagnetic energy in the amount hc/λ_1 is emitted as a quantum of radiation. He continues: "[The electron] then again moves away from the center 0 thanks to its remaining kinetic energy; however, due to the loss of energy, it cannot return to its original separation from the center but turns back at a smaller distance. During the period of greatest separation, it again undergoes an acceleration and during this period emits electromagnetic energy in the amount hc/λ_2. This is followed by another approach and emission of radiation in the amount hc/λ_2 etc. Thus, each time the electron under consideration falls on the solidly drawn curve,* it emits, in sequence, the wave lengths $\lambda'_1, \lambda_1, \lambda'_2, \lambda_2, \lambda'_3, \lambda_3 \ldots$."[27]

But Stark was unable to provide values for the energy loss of the electron at each turning point; nor did he in any way classify the various elliptical paths traversed successively by the electron in terms of energy. What he wanted to demonstrate (but simply states as assertion) was that the energy losses would become continually smaller and that the radiated energy would be greater in the perihelion than in the aphelion: "It can easily be seen that the wavelengths radiated at the aphelion are longer than those radiated at the perihelion; in addition, it can easily be shown that

$$\lambda_1 < \lambda_2 < \lambda_3 \cdots \text{ and } \lambda'_1, > \lambda'_2 > \lambda'_3 \ldots "$$

In this way, he felt that he had provided an explanation for the line bands that fade off at one "edge" toward the infrared region and at the other toward the ultraviolet. Actually, the mechanism of light emission described by him (not without contradictions) has nothing to do with band spectra.

To these ideas which he had published in 1908, he came back somewhat more cautiously in the preparation of the second volume of his *Atomdynamik*. As a result of this book, Stark's views bore fruit in a very surprising manner (see p. 155.)

5. X rays and the Light-Quantum Hypothesis

In 1909, Stark considered x rays in two papers published in the *Physikalische Zeitschrift;* in particular, he described how they are generated at the anode by deceleration of cathode-ray electrons. These studies led to extensive discussions with Arnold Sommerfeld who, late in 1909, was working on similar problems.[28]

Stark made remarkably accurate observations and presented views that confirmed the light-quantum hypothesis on which they were based. Thus he wrote: "The second fact whose significance with regard to radiation theory I pointed out recently is the ob-

*Stark drew a continuous spiral-like curve. Each successive revolution around the center is similar to an elliptical path.

servation that x rays cause emission of cathode rays even if far removed from their emission center, cathode rays whose energy is of the order of that of a single primary cathode-ray electron which itself emits x rays."[29] According to classical theory, the energy of x radiation decreases with radius as $1/r^2$ and the concentration of the energy within a small space, as in the case of the photoelectric effect, remains a complete mystery.

But both of Stark's papers also contain a fundamental physical error that shows his lack of theoretical insight which Arnold Sommerfeld pointed out in a letter dated December 10, 1909: ". . . I am not interested in engaging you in a dispute. At best, it would be unfair; for you excel in experimental ideas while I possess greater theoretical insight."[30]

Stark had made a comparison between the light-quantum hypothesis, in which he firmly believed, and a classical alternative in the form of the absurd assumption of bremsstrahlung having equal intensity in all directions. He could easily reduce this assumption ad absurdum by using the impulse equation, and he felt that he had thereby refuted classical theory and confirmed the light-quantum hypothesis; of course, he actually had only proved that bremsstrahlung cannot be spatially isotropic (as was already generally known). In fact, as Sommerfeld showed late in 1909, even classical electromagnetic theory provides an explanation for the origin and most of the significant properties of bremsstrahlung.

Thus, the "proof" offered by Stark for the light-quantum hypothesis, namely the anisotropy of bremsstrahlung, was irrelevant. A letter from Sommerfeld dated December 4, 1909, is quite clear in this regard: "If a recognized scientist like you commits such a serious error, he should straighten out the matter immediately in the interest of his less discerning readers . . ."[31]

The tone of Stark's subsequent letters to Sommerfeld showed signs of increasing irritation. His unfortunate argumentation was a great disservice both to his own views and to quantum theory.

Stark rendered an even greater disservice to himself by his immoderate polemic. While the earlier relationship between him and Sommerfeld had been quite good (as demonstrated for example by Sommerfeld's emphatic recommendation of Stark for a professorship at Aachen), it was now transformed into an enmity which would cast a shadow over the lives of both men.[32] The result of these disputes was that Arnold Sommerfeld and his colleagues were strengthened in their "confidence in the validity of electromagnetic theory even when applied to elementary processes."[33]

Against the numerous objections that were raised, Johannes Stark developed ideas to support the light-quantum hypothesis which were primarily intended to explain interference phenomena from the corpuscular point of view. His model illustrates the power of Stark's imagination but is just as characteristic for his disregard for theoretical arguments: He conceived a light-quantum field analogous to the microcrystalline structure of solids; here, without regard for theory, he conceived of strong forces of interaction ("ponderomotive forces") between individual light quanta. These would then close together as needed into "ordered light-quanta aggregates."[34]

Stark expressed these ideas in his correspondence with Lorentz in 1910; his reply to a detailed exposition of Lorentz's doubts was: "Your objections . . . impel me to undertake the planned experimental investigations immediately after the return from my trip."[35]

But Stark's interference measurements were "completely independent of the quantum hypothesis" as Lorentz pointed out in his letter of October 20, 1910. Lorentz continued: "Naturally, not everybody thinks in the same way about new and bold ideas; this depends on personal characteristics and probably also on age. I will certainly grant the great heuristic values of your ideas . . . but you must also permit me to test these ideas against well-known phenomena. Should present conceptions be overthrown, then you certainly would not demand that they be abandoned without defense. A new theory, if it is viable, gains strength from the objections which it provokes."[35]

References

1. Armin Hermann, "Hendrik Antoon Lorentz–Praeceptor Physicae. Sein Briefwechsel mit dem deutschen Nobelpreisträger Johannes Stark." In: *Janus*, vol. 53, 1966, pp. 99–114.

2. Johannes Stark, *Physikalische Zeitschrift*, vol. 9, 1908, pp. 767–773.

3. Ibid., vol. 8, 1907, pp. 881–884.

4. Ibid., vol. 8, 1907, p. 914.

5. Willy Wien, *Nachrichten der Göttinger Akademie*, vol. 1907, pp. 598–601.

6. Johannes Stark, *Physikalische Zeitschrift,* vol. 8, 1907, p. 882.

7. William Duane and Franklin L. Hunt, *Physical Review*, vol. 6, 1915, pp. 166–172.

8. Max von Laue, Geschichte der Physik, Frankfurt 1959, p. 147.

9. Hendrik Antoon Lorentz, *Physikalische Zeitschrift*, vol. 11, 1910, p. 352.

10. Armin Hermann, Albert Einstein und Johannes Stark. Briefwechsel . . . In: Sudhoffs Archiv, vol. 50, 1966, pp. 267–285.

11. Ref. 2, p. 772.

12. Ernst Gehrcke and Rudolf Seeliger, *Verhandlungen der Deutschen Physikalischen Gesellschaft*, vol. 14, 1912, pp. 335–343 and 1023–1031.

13. James Franck and Gustav Hertz, Die Elektronenstossversuche (*Dokumente der Naturwissenschaft*, vol. 9), Munich 1967, p. 13.

14. Ibid., p. 36.

15. Ibid., p. 47.

16. Gustav Hertz, Max Planck zum Gedenken, Berlin 1959, p. 39. Gerald Holton has already emphasized that the Franck-Hertz experiment was not carried out to confirm Bohr's theory. See Gerald Holton, On the Recent Past of Physics, *American Journal of Physics*, vol. 29, 1961, p. 808.

17. Niels Bohr, *Philosophical Magazine,* vol. 30, 1915, p. 404.

18. Johannes Stark, *Physikalische Zeitschrift*, vol, 9, 1908, p. 892.

19. Albert Einstein, *Annalen der Physik*, vol. 37, 1912, pp. 832–838.

20. Johannes Stark, *Annalen der Physik*, vol. 38, 1912, p. 467.

21. Ref. 19, vol. 38, 1912, p. 888.

22. Albert Einstein, letter to Emil Warburg, Autographenkatalog Stargardt, No. 574, April 25, 1912 (or 1911).

23. Johannes Stark and Paul S. Epstein, Der Stark-Effekt, (*Dokumente der Naturwissenschaft*, vol. 6), Stuttgart 1965, p. 11.

24. Johannes Stark, *Physikalische Zeitschrift*, vol. 9, 1908, p. 86.

25. Johannes Stark, unpublished manuscript, written 1944/47, Stark estate, Staatsbibliothek Preussischer Kulturbesitz Berlin.

26. Paul S. Epstein, letter to Johannes Stark, Stark estate. June 19, 1916.

27. Ref. 24, p. 88.

28. Armin Hermann, Die frühe Diskussion zwischen Stark und Sommerfeld über die Quantenhypothese [Correspondence Stark/Sommerfeld]. In: *Centaurus*, vol. 12, 1967, pp. 38–59.

29. Johannes Stark, *Physikalische Zeitschrift*, vol. 10, 1919, p. 912.

30. Ref. 28, p. 49.

31. Ibid., p. 45.

32. Ibid., p. 53.

33. Arnold Sommerfeld, *Physikalische Zeitschrift*, vol. 10, 1909, p. 976.

34. Ref. 1, p. 105.

35. Ibid., p. 109.

5

Arthur Erich Haas   The First Application of
                                         Quantum Theory to the Atom
                                         (1910)

1. Atomic Models

After the discovery of the electron it became known as a fundamental fact of great importance that the same type of particles
always appears, regardless of whether they originate in incandes-
cent metals, from irradiation with ultraviolet light, or are generated by (primary) cathode rays: "Since corpuscles similar in all
respects may be obtained from very different agents and materials, and since the mass of the corpuscles is less than that of
any known atom, we see that the corpuscle must be a constituent
of the atom of many different substances."[1] Physicists were in
agreement on this conclusion and also that the electrons—or the
single electron?—are responsible for the emission of spectral lines
in the atom. No answer was forthcoming regarding how this actually takes place.

The following models of the atom were proposed at about the
same time or else followed each other in quick succession: The
elastically bound electron of Lorentz (1896), the dynamid concept of Lenard (1902), Thomson's electrically positive atomic
sphere (1903), the saturnian system of Nagoaka (1903), the magnetic atom of Ritz (1908) and, finally, the Rutherford model
(1911).

The most convincing of these was the atomic model of Thomson;
it was used, for example, by Max Born in his 1909 Göttingen
habilitation lecture.[2] At a Munich physics colloquium on
November 22, 1911, Peter Paul Ewald, then a student of Sommerfeld, demonstrated the Thomson atomic concept by means of
magnetic models.

Still, the majority of physicists was not convinced by any of the
proposed models. One basic uncertainty, indeed, was that no one
knew exactly how many electrons are responsible for a line
spectrum; also, no information was available on the total number
of electrons in the atom. In most cases, as we know today, too

many attributes were sought: in addition to the spectral lines and the regularities of the Periodic Table of the Elements, radioactive phenomena were also to be explained by the model.[3]

These early attempts show a lack of fundamental prerequisites necessary for a successful treatment of the problem; it seems to us that a clarification of many concepts on a "prequantum-theoretical" basis was missing.* The needed clarification was essentially provided by Bohr in 1912/13, an achievement which is often underestimated.

On October 14, 1904, Max Abraham had this to say: "The spectral lines can hardly be explained on the basis of electron mechanics alone. It will probably be necessary to add a new fundamental hypothesis concerning the interaction of matter (i.e., of positive electricity) with the electrons. It is important that this be established in a satisfactory manner. I doubt whether much progress will be made along this line during the next few years."[5]

The widespread and continued ignorance concerning the mechanism of line emission is clearly shown in a letter from Friedrich Paschen to Sommerfeld of March 1913: "I [Paschen] am of the opinion that the theory of Ritz is a good starting point for a treatment of the problem because it stipulates the vibration of only a single electron, and the doublet is represented by two magnetic fields which might well be influenced by an external field. Einstein is in agreement with this. However, initially he did not share my opinion of Ritz's theory, with which he is not too familiar. Only after I had told him what I have just stated did he agree that the single electron was very important. In Lorentz's

---

*For example, the concept of atomic number and the various "internal states" as well as the separation of the atom into a "nuclear" region and an "atomic" region as such. Friedrich Hund expressed the opinion[4] that "the whole of quantum mechanics might have evolved from the laws of spectral series" ("without Planck and Einstein, without considering harmonic oscillators, light quanta, and specific heats . . . and also without Rutherford"). The present author feels that this may well be a logical but not a historical possibility.

and Voigt's theories we can assume several coupled electrons in order to explain the anomaly."[6]

2. Planck's Resonators

In his derivation of the radiation formula of 1900, Planck had assumed linear resonators as the response to electromagnetic radiation. But Planck had provided these resonators with only very formal properties.

Planck's undamped resonators act only over a very narrow range of frequencies $\nu \pm d\nu$; in other words, they can only emit and absorb a definite fixed frequency. Thus, they are not able to transform a partial wave of frequency ν_1 into a partial wave of frequency ν_2. As Willy Wien and Henri Poincaré remarked in discussing Planck's lecture at the first Solvay Congress, an interaction of the resonators with each other must therefore be assumed, in addition to the interaction between radiation and the resonators. Of course, Planck implicitly talked of such an interaction from the start but explained the matter no further at the turn of the century. In response to direct questioning in Brussels in 1911, he said: "Electrons might be the energy transfer agents."*[7]

Planck was satisfied with the statement that the type of electromagnetic resonators is not relevant for the derivation of the radiation formula. It is characteristic of his approach that he did not attempt to draw conclusions regarding the properties of atoms from those which he had determined for the electromagnetic oscillators.

This is quite different from the actions of other experimental physicists such as Otto Lummer, Joseph John Thomson, Ernest Rutherford, and Willy Wien. Thus, for example, Wien in 1909

---

*In 1905, Albert Einstein assumed three different structures: in addition to the electromagnetic partial waves and the Hertzian oscillators, there were the "freely moving molecules" which provide the energy exchange of the resonators among each other by means of collisions with the resonators. However, the energy of the "free molecules" is not taken into consideration.

viewed the electromagnetic resonators as real atoms which in addition to their ability to absorb and emit radiation energy also possess other characteristics; he considered them as structures which can be ionized by ultraviolet light or by x rays, for example.[8]

3. Arthur Erich Haas

Haas was the first to attempt to apply Planck's quantum of action to the constitution of the atom. It is quite clear who provided the impetus for this work and what considerations impelled him.

Early in 1909 Einstein had become convinced that a relationship must exist between two missing aspects of Maxwell's theory (which can explain neither the quantum nature of radiation nor the existence of electrons): "It seems to me that we can conclude from $h = e^2/c$ that the same modification of theory that contains the elementary quantum e will therefore also contain as a consequence the quantum structure of radiation."[9]

Einstein's paper, which appeared in the *Physikalische Zeitschrift* on March 15, 1909, provoked Willy Wien to respond in an article written in the spring of 1909 on the "Theory of Radiation" for the *Encyclopädie der mathematischen Wissenschaften*: "The opinion expressed by Einstein . . . that the magnitude of the element of energy is related to the magnitude of the elementary quantum of electricity is unacceptable to me for the present because the element of energy, if it has any physical meaning at all, can probably be derived only from a universal property of atoms."[8]

Arthur Erich Haas, in turn, carefully read Wien's paper. Section 2 of volume V,3 of the Mathematical Encyclopedia was published on September 28, 1909, and just at that time, Haas was studying the "most recent literature of physics"[10] in order to find a theme for a thesis required by the Philosophical Faculty of the University of Vienna.

4. The 1910 Theoretical Attempts

Early in 1909, Haas, interested and gifted in many fields and 25 years old at that time, had submitted to the philosophical faculty of the University of Vienna his request for a university appointment (as Privatdozent) in the field of history of physics and at the same time had submitted as his thesis for habilitation his recently published study "The History of the Principle of Energy Conservation."

As he pointed out in his memoirs, the unusual situation of a historical dissertation presented a dilemma to the physicists primarily responsible for its evaluation, Franz Exner and Victor von Lang. Thus, after many months, they reached the "not ill-intentioned decision" that Haas would first have to prepare a purely physical, not historical, study which he should then submit together with his original dissertation.

Dismayed by this decision, Haas returned to his earlier law studies, which had been his secondary field of interest. After passing the bar examination, Haas returned to the field of physics with the intention of meeting the requirements of the philosophical faculty.

He now turned to his study of the most recent literature of physics, that is, the work of Planck, Rayleigh, Jeans, and Lorentz concerning thermal radiation; he apparently found his way into these problems through the encyclopedia article by Willy Wien, which he immediately followed up by a study of Planck's book, "Lectures on the Theory of Thermal Radiation."[11] He noted that Planck's theory alone was really capable of deriving the experimentally determined spectral distribution for all frequencies. "And yet, as its author repeatedly points out, Planck's theory can by no means be considered complete. For one of its essential requirements is a universal constant, to which Planck has given the name of 'elementary quantum of action,' whose physical meaning is still completely undefined; and, as Planck explains, our understanding of the thermodynamics of radiation will not be complete until this 'constant h' is clearly understood in its universal and, in particular, its electrodynamic significance."[12]

In Planck's view, in order to derive the normal distribution of thermal radiation it is necessary to go back to the emission and absorption processes of radiation. This drew Haas's attention to "optical resonators" in the radiation field. From reading J. J. Thomson's book *Electricity and Matter*, whose central theme is the problem of the constitution of the atom, Haas was led to substitute real atoms for the idealized Hertzian oscillators used by Planck.

"In the following investigation," wrote Haas in his paper of February 1910, "Thomson's model of a hydrogen atom will be examined as a special case of an optical resonator." This clearly demonstrated Haas's intention to relate this question of the nature of the quantum of action to that of atomic structure: "The present study will attempt to provide an electrodynamic interpretation of Planck's element of action. Its goal will be the development of an equation relating the constant h to the fundamental quantities of electron theory . . . which would also provide the basis for a more exact evaluation of the radius of the hydrogen atom."[12]

To be sure, as might be expected, Haas's approach is strictly "classical"; his efforts constitute one of the many attempts to derive the relation $\epsilon = h \cdot \nu$ or the nature of the quantum of action from known principles, or at least to fit this law into known principles. But since he undertook his considerations starting with the problem of atomic constitution, his attempt was extraordinarily significant.

The atomic model of Thomson which Haas used initially makes no specific statement regarding distribution of the positive charge; thus, this model is more general than the planetary atomic model of Rutherford, which it includes as a special case. Haas now gave particular attention to the circular orbits of electrons at the surface of the positively charged sphere; but since the forces acting at and outside the surface of a charge distribution with spherical symmetry are exactly the same as those that would exist if the total charge were concentrated at the center, the size of the nucleus is immaterial. Thus, Haas's calculations are based on an atomic model which is equivalent to that of Rutherford.

As we know from classical physics, the expression that equates centrifugal force to Coulomb attraction relates the radius of the orbital path to the orbital frequency; in order to determine specific radii, a second condition is required.

Haas assumed as a matter of course that the motion of the electrons takes place within, or at most on the surface of, the positively charged sphere; when an electron attains greater distance, it is lost. He now looked at the potential and kinetic energy for circular electronic orbits as a function of radius. Both energies increase from $r = 0$ (center of sphere) to $r = a$ (surface of the sphere); the kinetic energy then decreases (according to $1/r$) for $r > a$, while, of course, the potential energy continues to increase as $r \longrightarrow \infty$, since an attractive force continues to act on the electron even in the external region. (The corresponding equation of Haas[13] is in error, as is his related assertion that the potential energy would again decrease with increasing r for $r > a$. If we exclude the case $r > a$, we can agree with Haas on a maximum potential energy for $r = a$.)

Haas developed a fortunate thought in comparing the potential energy of the atomic electron e^2/a with Planck's element of energy $h\nu$. Since the magnitude of the atomic radius was unknown and the elementary charge and the quantum of action were only known approximately, Haas could merely estimate the orders of magnitude. He found that both energies had values of the same order of magnitude. "Thus, it is not only quite reasonable but also substantiated by the quantitative relationships, for us to assert the hypothesis that, in a state which is particularly characterized by its unusual simplicity (for the case of $r = a$), the energy of the hydrogen atom can also be described by the simplest possible expression, namely the quantity $h\nu$; therefore the element of energy which appears in the hydrogen atom as an optical resonator is identical with its maximum energy."[14]

Haas now wrote

$$|E_{pot}| = h\nu. \tag{1}$$

This is the decisive quantum-theoretical condition which, together with the classical relation equating centrifugal force with coulomb attraction,

$$m\omega^2 a = e^2/a^2, \tag{2}$$

determines the radius a and the angular velocity ω of the electron. Viewed historically, we can almost take as a matter of course that Haas directly identifies the frequency ν of equation (1) with the orbital frequency of the electron: $\omega = 2\pi\nu$.

Haas's equation applies to the ground state of the hydrogen atom, in agreement with Bohr's subsequent condition; Haas therefore obtained the correct "Bohr" radius a of the hydrogen atom:

$$h = 2\pi e \sqrt{am}, \tag{3}$$

which, solved for a, yields

$$a = h^2/4\pi^2 e^2 m. \tag{4}$$

Characteristically, Haas wrote down only equation (3) and not the solution (4) for a, since he considered the dimensions of the atom to be fundamental quantities from which Planck's quantum of action h could then be derived.

Haas's quantum condition (1) only provides the ground state of the atom; missing are the decisive excited states and of course the transitions which are the essence of Bohr's theory and which, in conjunction with spectroscopic findings, led to Bohr's epoch-making success. With his substitution of real atoms for the Hertzian oscillators used by Planck, Haas should logically have considered higher levels of excitation. Of course, he lacked all the information necessary for a casual solution of such a difficult problem.

Haas used Balmer's equation, though only in a very summary fashion. He supplemented his equation (1) by a second hypothesis, namely, that the frequency ν derived from (1) is in agreement with the constant $\nu_\infty = 8.23 \times 10^{14}$ $\sec^{-1}$ of Balmer's equation which, according to Haas, (and rewritten for frequencies) is

$$\nu = \nu_\infty (1 - 4/n^2)$$ (5)

Thus, Haas's constant ν_∞ corresponds to our present-day definition $R/4$ (R is the Rydberg frequency $R/c = 109737.41$ cm^{-1}) which follows directly from (5) or from a comparison of the values for $R/4$ and ν_∞.

Since the frequency ν_∞ is known to three decimal places, Haas now intended to undertake a more accurate determination of the two constants, the electrical elementary quantum and the atomic radius. But Haas's second hypothesis

$$h\nu_\infty = |E_{pot}|$$ (6)

is in error by a numerical factor of 8.*

If Haas's equation, which interrelates the charge of the electron, its mass, Planck's constant, and the Rydberg constant,[15] is solved for ν, the result is

$$\nu_\infty = 4\pi^2 \, me^4 / h^3.$$ (7)

The correct equation would have been

$$R = 4\nu_\infty = 2\pi^2 \, me^4 / h^3$$ (8)

Thus, the difference with respect to (8) is not, as Haas later claimed, due to application of Thomson's instead of Rutherford's model of the atom. The discrepancy of a factor of 8 could

*Bohr uses the Rydberg frequency as the limiting frequency emitted in the transition from $n = \infty$ to the ground state $m = 1$: $h \cdot R = |W_1| = |E_{pot}|/2$.

have escaped Haas's attention, since the pertinent constants were only known with considerable uncertainty at that time.

5. First Reactions to Haas's Ideas

The reaction provoked by his thoughts among the physicists of Vienna is vividly described by Haas in his autobiography: "The fact that my achievement is generally recognized today only increased the pain I experienced throughout my later life because of the fact that the narrowmindedness of the established leaders in physics stifled my initial attempts which might have led to great and possibly fundamental achievements. ... In Vienna ... I was at first met only by disdain and even derision. When I lectured to the Chemical-Physical Society of Vienna concerning my studies, Lecher thought he was particularly witty when he referred to the presentation during open discussion as a carnival joke,* and when Laurenz Müllner questioned [Friedrich] Hasenöhrl concerning me, he was told that I could not be taken seriously since I naively mixed up scientific fields which were completely unrelated, such as quantum theory (as something thermodynamic) and spectroscopy (as something optical). Understandably, I became completely discouraged."[16]

However, Hasenöhrl, the young professor of theoretical physics at the University of Vienna, did not persist for very long in his strong condemnation of Haas's work. He began to realize that some of his other colleagues, particularly Albert Einstein and Johannes Stark, were developing quite extraordinary physical concepts with regard to Planck's quantum law, $\epsilon = h\nu$. Thus, Hasenöhrl was led to revise his judgment: he encouraged Haas to resubmit his request for a university appointment (as Privatdozent) in History of Physics, in accordance with the original decision of the philosophical faculty; however, Haas was asked first to submit another purely physical paper.

The opinions of the Leipzig physicists Otto Wiener and Theodor

*The lecture was held in February 1910, carnival time in Austria.

Des Coudres underwent a change similar to that of Hasenöhrl.* In a letter to Haas of November 2, 1911, Sudhoff informed Haas that the quantum approach, which had previously shocked some physics professors in Leipzig, "was now considered to be quite harmless."[17] That was the extent of the progress of his ideas among the Leipzig physicists, with which Haas was expected to be satisfied after a year and a half after his publication.

However, it was not long before the significance of the relationship between the atomic quantities e, m, a, and Planck's quantum of action h expressed by Haas's equations (3) and (4) was recognized.

During a series of six lectures presented by Lorentz between October 24 and 29, 1910, in Göttingen, he referred to Haas's equation as "a daring hypothesis." But he emphasized that Haas's equation (3) at least expressed the correct order of magnitude.[18] Sommerfeld, who also had become familiar with the work of Haas, picked up the fundamental idea of the relationship between Planck's quantum of action and the atomic quantities. But at the 83rd conference of physical scientists in Karlsruhe in 1911 he took the opposing viewpoint "not to explain h on the basis of molecular dimensions but rather to view the existence of the molecule as a function and result of the existence of an elementary quantum of action."[19] This idea came to fruition with the establishment of Bohr's theory. Today, when we have probably reached a point in the theory of elementary particles that corresponds to the state of atomic physics in 1911, we feel that the introduction of an "elementary length" will be a decisive step forward.

Within the small, exclusive circle of physicists at the first Solvay Congress in Brussels, detailed discussions were held concerning Haas's atomic model and equation (3), particularly in conjunction

---

*Karl Sudhoff, who in 1906 had founded the first Institute of the History of Medicine in Leipzig, pursued a plan to call Haas to his institute as associate professor. The judgment of the Leipzig physicists was therefore decisive.

with a review of fundamental considerations presented by Planck. Sommerfeld, for example, remarked: "With respect to Haas's hypothesis I would like to remark that the numerical relationship between the quantity h and the molecular dimensions also follows from the much more general action principle without such special requirements. . . . But in this instance the relationship is limited to a definite region of ultraviolet vibrations and thus does not appear to be of fundamental significance. I, for one, would prefer a general hypothesis for the quantity h to the special atomic models."[20]

To this, Lorentz replied: "Herr Sommerfeld does not deny a connection between the constant h and the atomic dimensions. This can be expressed in two ways: Either the constant h is determined by these dimensions (Haas), or the dimensions, which are ascribed to atoms, depend on the magnitude of h. I see no great difference between these views."[21]

When Arnold Sommerfeld congratulated Niels Bohr on his calculation of the Rydberg constant by way of a postcard dated September 4, 1913, he reiterated his skeptical view "regarding atomic models as such."[22] As Léon Rosenfeld points out, this opinion of Sommerfeld amused Bohr in later years when, in the 1920's, it was Sommerfeld who, as one of the leading physicists concerned with the problems of quantum theory, particularly wanted to maintain the Bohr model of the atom.

The linking of the Rydberg constant to the physical constants of electron mass, elementary charge, and quantum of action (which, however, was in error by a factor of 8) was not well received. This is peculiar; for when Arnold Sommerfeld on September 4 and William C. Oseen on November 11, 1913, congratulated Niels Bohr on the achievement of his atomic theory, they specifically emphasized this calculation of the Rydberg constant by Bohr.

Haas's great accomplishment lay in establishing a relationship between the quantum of action and atomic dimensions; it was relatively unimportant that his equation (1) be exactly correct, for the Bohr (or, historically more correct, the Haas) atomic radius a is not an observable quantity. The connection between Haas's model and the Rydberg constant does not involve the application of any spectroscopic findings. In fact, Haas completely neglected consideration of the emission and absorption of radiation which Planck had considered the essential property of his Hertzian oscillators.

Today we would say that Haas concerned himself solely with the ground state of the atom. Historically, it is more correct to say that Haas considered the atom as such, neglecting all changes this atom might undergo (emission and absorption of light, ionization and recombination).

6. Quantum-Theoretical Atomic Models of 1911

Early in 1911, Arthur Schidlof attempted to extend Haas's model with respect to the emission and absorption of light. But since the concept of various "internal states" of the atom was not available to him (Bohr was the first to develop it late in 1912), Schidlof, like others, was forced to establish a causal relationship between the emission and absorption of light with the absorption and emission of electrons. The energy principle cannot be satisfied in any other way. Planck also mentioned this widely held view in a letter to Sommerfeld: "As regards the emission of electromagnetic radiation, I am not opposed to the view that this is always the result of [emitted or absorbed] electrons; but if this radiation is monochromatic, then the emission must occur in energy quanta $h\nu$."[23]

Schidlof considered light emission as the absorption of electrons with zero kinetic energy; the increased potential energy is converted into light energy: "We now assume that radiation energy is absorbed only if an electron is removed from the positively charged sphere under the influence of the external electromagnetic field. Conversely, the emission of radiant energy is only possible with the absorption of an external electron."[24] It is clear that such a mechanism could not be successfully incorporated into a quantum-theoretical atomic theory and that Schidlof's efforts were destined to fail.

Friedrich Hasenöhrl was not able to take Haas's proposal seriously early in 1910 for the reasons quoted in Haas's autobiography (see p. 96). But by the time of a convention of physical scientists in Karlsruhe in September 1911, he also advanced quantum equations toward an understanding of atomic structure.* Hasenöhrl attempted nothing less than to derive Deslandre's law of band spectra and Balmer's formula for spectral series.[25]

In a very general way, starting with Planck's resonator, Hasenöhrl looked for an oscillator better suited to describe the behavior of a real atom. This oscillator is then also to be considered as a quantum-mechanical system: "But it would not be consistent also to assume here equally large energy elements; it is more in keeping with Planck's view to divide the phase space (the phase plane) into equally large elementary regions h bounded by curves of constant energy."[25]

This seemingly worthwhile effort was destined to fail because Hasenöhrl did not recognize the empirical equations of band and series spectra as the differences of terms corresponding to atomic states. A more thorough treatment of the problem on the same basis as that chosen by Hasenöhrl would probably have been successful.

As significant progress made in 1911, we can record these initial attempts to understand the properties of atoms and molecules by means of the quantum concept. Phrased another way: The oscillator conceived by Planck in 1900, a step which signaled the beginning of quantum theory, had now been transferred from a formal concept to a physical reality.

Arthur Schidlof noted: "We have based our calculations on the

*Hasenöhrl's correspondence has been missing since his early death on October 7, 1915, during the First World War. We do not know what caused Hasenöhrl's change of mind in 1911 and can merely suggest the general trend of thinking at the time: 1910 saw the beginning of a universal shift of opinion in favor of the quantum concept.

supposition that atoms can behave as Planckian resonators. But it should be kept in mind that such resonators are generally not atoms. The resonant vibrations of atoms lie far in the ultra-violet region of the spectrum. Those resonators which react to visible light already have significantly greater radii than atoms. . . . The constitution of these resonators must, by the way, be exactly the same as that of atoms."[26]

Willy Wien, who in 1909 had already drawn a parallel between the properties of Planck's resonators and those of real atoms,[8] also emphasized their difference in the same year, 1911: "The Planckian resonator lacks one essential property of real molecules, namely the capacity to change the wavelength of radiation. . . "[27]

It is instructive to compare the ideas of Haas, Schidlof, Hasen-öhrl, and Nicholson with Bohr's approach; the comparison shows that a mere resolve to apply a quantum concept in order to explain the nature of the atom was not yet enough. It was still necessary to provide new ideas and information concerning the physical characteristics of atoms.

References

1. Joseph John Thomson, Electricity and Matter, New Haven 1904, new edition 1912, p. 90.

2. Max Born, *Physikalische Zeitschrift*, vol. 10, 1909, pp. 1031–1034.

3. Ref. 1, p. 92.

4. Friedrich Hund, Geschichte der Quantentheorie, Mannheim 1967, p. 68.

5. Max Abraham, letter to Arnold Sommerfeld, Sommerfeld estate, Stuttgart University, October 14, 1904.

6. Friedrich Paschen, letter to Arnold Sommerfeld, Sommerfeld estate, March 1913.

7. Die Theorie der Strahlung und der Quanten, Arnold Eucken, ed. [Transactions of the 1st Solvay Congress], Halle 1913, p. 105.

8. Willy Wien, Encyklopädie der mathematischen Wissenschaften, vol. V, 3. Leipzig 1909–1926, p. 356.

9. Albert Einstein, *Physikalische Zeitschrift*, vol. 10, 1909, p. 192.

10. Arthur Erich Haas, Der erste Quantenansatz für das Atom (Dokumente der Naturwissenschaft, vol. 10), Stuttgart 1965, p. 9.

11. Max Planck, Vorlesungen über die Theorie der Wärmestrahlung, Leipzig 1906.

12. Ref. 10, p. 28.

13. Ibid., p. 55.

14. Ibid., p. 58.

15. Ibid., p. 49.

16. Ibid., p. 16.

17. Ibid., p. 17.

18. Hendrik Antoon Lorentz, Collected Papers, vol. 7, The Hague 1934, p. 245.

19. Arnold Sommerfeld, Physikalische Zeitschrift, vol. 12, 1911, p. 1066.

20. Ref. 7, p. 102.

21. Ibid., p. 103.

22. Léon Rosenfeld, in: Niels Bohr, On the Constitution of Atoms and Molecules, Copenhagen and New York 1963, p. lii.

23. Max Planck, letter to Arnold Sommerfeld, Sommerfeld estate, Stuttgart University, April 6, 1911.

24. Arthur Schidlof, Annalen der Physik, vol. 35, 1911, p. 93.

25. Friedrich Hasenöhrl, Physikalische Zeitschrift, vol. 12, 1911, p. 933.

26. Ref. 24, p. 99.

27. Ref. 7, p. 99.

6

Arnold Sommerfeld   Interactions between
Electrons and Molecules
(1910–1912)

1. Skeptical Reticence

On April 1, 1900, Arnold Sommerfeld was appointed full professor for applied mechanics at the Technical University of
Aachen. His interest was directed, like the efforts of his adored
teacher and friend Felix Klein, toward the application of mathematical methods to technical problems, thereby providing a bridge
between mathematics and technology.[1]

Only after Sommerfeld was called to Munich in the fall of 1906
to assume his professorship in theoretical physics, was he forced
by the requirements of the academic curriculum to familiarize
himself with the problems of modern physics. As in Aachen, he
accomplished this primarily by engaging his assistants and students
(in the early years, Peter Debye in particular) in conversations
about ideas of current interest to him.

The following verbatim description was given by Debye:[2]
"What he did was, he invited us to come to his house. We came to
his house in the evening at 8 o'clock, had the evening meal, sup-
per. And then you sit in his room. And in his room he began to
talk. He asked you about it, although you did not know anything
about it. He tried it out, so to say. And in this way, I learned a lot.
But you sat there until 11 or 12 o'clock in the evening and you
talked and he talked. The sessions were perhaps twice a week or
so. It was not regular. You might meet him and he would say:
'Oh come up, I have to talk about something.' " When he transferred from Aachen to Munich, Sommerfeld took Debye along.
The latter relates: "Then he said: 'Well now I have to study in
order to give my courses.' And we studied together in order that
he could give his courses . . . And at the time he began also to talk
about radiation. And then Planck comes in and so on. But he had
to study that, and he studied that the way he had in Aachen, you
see. He talked to me. Then I had to sit there and make some
remarks."

Sommerfeld had a very ambitious view of what was required of him in his lectures. Thus, as a part of his academic program he turned his attention to quantum problems as well. A statement to this effect by Debye is in agreement with a letter from Sommerfeld to Stark dated October 10, 1908: "I would be very grateful to you for an extra copy of your quantum spectrogram* which I would like to use in my courses and above all in order to become definitely converted to Planck's fundamental hypothesis."[3]

In the same letter, Sommerfeld strongly urged Stark to endorse certain remarks made by Debye concerning the quantum interpretation of the Doppler effect in the course of discussions after a lecture delivered by Stark at the meeting of physical scientists in Cologne in 1908, for publication in the *Physikalische Zeitschrift*. In these remarks, Debye had somewhat extended the quantum interpretation of the Doppler effect given by Stark.[4] Sommerfeld reverted to this matter in a later letter dated December 16, 1909, which reads: "I have just noted a paper by Royds on the Doppler effect in the Philosophical Magazine, which credits the relationship $v \sqrt{\lambda} =$ const. to Stark and Steubing. This is probably due to the fact that in your definitive publication in the Annalen der Physik there is no mention of Debye. Perhaps you could correct this matter at a suitable moment; the brief communication in the *Physikalische Zeitschrift* can be easily overlooked. In any event, a talented newcomer like Debye should be encouraged."[5]

Late in 1909, Stark's publications concerning x-ray bremsstrahlung[6] provoked strong protests from Sommerfeld. The illogical contradictions of Stark's presentation proved to Sommerfeld that Stark lacked all theoretical insight. The alternative posed by Stark to the light-quantum hypothesis, which he defended, was the absurd assumption of a bremsstrahlung radiating with the same intensity in all directions (see p. 83).

In a letter addressed to Stark on December 4, 1909, Sommer-

*This concerns the "subdivided" light intensity of the Doppler effect, which was erroneously interpreted by Stark as a quantum effect (see p. 75).

feld wrote: "Your last paper on x rays has encouraged me to publish some ideas which I have considered for some time and which I have often discussed with Röntgen. Their experimental verification resulted in a doctoral thesis[7] which Röntgen presented to our institute . . . I hope you will become convinced that the bremstheory of x rays leads directly to the same results for which you call on the (still quite hypothetical and uncertain) light-quantum theory. I do not doubt the significance of the quantum of action but it seems to me and also to Planck that the way in which you have extended its application is quite risky."[8]

Sommerfeld's publication appeared in the *Physikalische Zeitschrift* on December 15, 1909. Its substance was a derivation of the angular dependence of bremsstrahlung intensity based on classical electromagnetic theory. Sommerfeld may have been particularly pleased with his calculations because they involved a relativistic effect: For electron velocities v which are small in comparison with the velocity of light, the maximum intensity is directed transversely. As v increases, the maximum moves from $\phi = \pi/2$ toward a smaller azimuth; here $\phi = 0$ corresponds to the direction of the (initial) velocity of the decelerated electrons.

These are the intensity differences in various directions that were experimentally determined by Stark. Sommerfeld stated that "the electromagnetic theory has long made provision for this"[9] and mentioned that "an indication of this lack of symmetry of intensity had already been given in the Munich dissertation of Herr Bassler." In the spring of 1912 Sommerfeld was still extremely interested in the question of directional distribution of x rays. Max von Laue, then a lecturer at Sommerfeld's institute, suggested in March of 1912 that x rays be passed through crystals in order to check the space-lattice hypothesis (which later led to the famous discovery of x-ray interferences). Von Laue discusses this in his autobiography: "The only difficulty was that Sommerfeld initially was not convinced by this idea and would have preferred to see an experiment on the directional distribution of radiation emanating from the anode."[10]

Sommerfeld was convinced by the success of his calculation; there was no room in x-ray bremsstrahlung for the quantum of action: "In the second 'fluorescence component' of radiation, an absorption and emission of energy occurs in the atom. It is quite possible that Planck's quantum of action plays a role here. I would like to see the wavelength calculation of x rays as proposed by J. Stark and W. Wien limited to this (generally most significant) portion of the radiation. However, it seems to me that the quantum of action has nothing to do with the first 'brems component' of the radiation. There, it would be better to continue to talk of 'impulse width' rather than of the 'wavelength' of x rays. An attempt should be made to explain the properties of this component entirely on the basis of electromagnetic theory."[9]

Sommerfeld closes his presentation with an expression of his conviction that the proven Maxwell equations rather than the speculative light-quantum hypothesis are the basis of the correct physical theory: "It further seems to me that this will again strengthen our confidence in the validity of the electromagnetic theory when applied to the elementary processes in electrical fields, a confidence which to some extent appears to have been shaken by the most recent speculations concerning light quanta."[11] For Max Planck as well as other physicists, the net result of the discussions between Stark and Sommerfeld appears to have been a strengthening of the wave theory of light and x rays at the expense of the light-quantum concept. In January 1910, Planck wrote in the *Annalen der Physik:* "I cannot recognize the experiments of J. Stark on x rays as convincing proof, any more than I can the deductions of A. Einstein in favor of the corpuscular theory of light at the present time. Such proof would require far more definite information concerning the nature of x rays as well as of the individual processes involved in their origin than is available at present. It should be especially stressed that, contrary to Stark's assumption, the wave theory of x rays certainly does not stipulate the same intensity and frequency of radiation in all directions; in each case, this depends on the type of brems mechanism of the electrons."[12]

But even these discussions with Stark, which had strengthened Sommerfeld's confidence in the validity of classical electromagnetic theory, did not cause the latter to reject all application of quantum concepts. This is shown, for example, by his letter to Stark of December 16, 1909, quoted on page 104.

2. A Change of Philosophy

Early in 1910 Sommerfeld was still quite skeptical with respect to the quantum hypothesis; we know that later, after the development of Bohr's model of the atom, he became an energetic defender of quantum theory. Is it possible to find out when and under what influences Sommerfeld changed his philosophy?

He presented his quantum ideas in September 1911, at the congress of German physical scientists and physicians in Karlsruhe and again early in November of the same year at the Solvay Congress in Brussels. The thoughts developed there must have been conceived several months previously. As a matter of fact, in a paper concerning the structure of γ rays, presented on January 7, 1911, to the Bayerische Akademie,[13] we find the nascent idea for the fundamental concept of both of his important lectures presented later that year.

Thus, the decisive time span during which Sommerfeld became a convert must lie between February and December 1910. Fortunately, it has been possible to locate a letter from Debye to Sommerfeld dated March 2, 1910.* At that time, Sommerfeld was in Göttingen for a visit and Debye wrote him four pages which he introduced with the following words: "I have now reached a definite point of view in connection with my thoughts concerning radiation, and since I feel that there are some aspects which might

____

*Since Debye was Sommerfeld's assistant in Munich until April 1, 1911, they were generally not dependent on correspondence for exchanges of ideas.

also be of interest to you in Göttingen, I want to bring this matter to your attention . . ."[14]

This sounds as if the young Debye was influencing his respected teacher Sommerfeld in favor of the quantum theory. During his interview for the "Sources for History of Quantum Physics" Debye understandably no longer remembered that particular letter. But the general impression which he gave of his relationship to Sommerfeld agrees completely with it.

Kuhn: "I take it that at this stage of the game you were perhaps even further than he was?" *Debye:* "Oh yes, that is true. He had not paid very much attention to it, and I was interested in them [the quantum-riddles]. I was going to give this course and so I went into it. And then I talked to him."[2]

Thus we can note that Debye had a significant influence on Sommerfeld. We also believe that a (possibly indirect) influence by Walther Nernst may have been at work. In late 1909 or early 1910, Nernst had become a vigorous proponent of the quantum concept and by June 1910 began to prepare himself for the quantum conference to be held in Brussels. It is quite certain that Sommerfeld, who was always extremely well informed about all efforts made in his field of interest (a fact which was consistently emphasized by his students) had early information on Nernst's intentions. Sommerfeld's wish to stand always in the forefront of scientific research may have been a significant factor in causing him to turn seriously to the quantum problem.

What was the influence of Max Planck in this regard? The two physicists met at the conference of physical scientists in Salzburg in September 1909 where they attended the sensational lecture by Einstein concerning the theory of relativity and the quantum concept. (Planck and Sommerfeld had previously met at a conference of physical scientists in Stuttgart in 1906. Planck erroneously recalled later that his first meeting with Sommerfeld had taken place in September 1910 in Königsberg.) It can be assumed that, while in Salzburg, they discussed Einstein's ideas in particular and also the quantum problem in general.

However, it is unlikely that a strong influence would have been exerted by the circumspect Planck. In his letter to Stark written on December 4, 1909, ten weeks after the Salzburg meeting, as quoted on p. 105, Sommerfeld stated that he and Planck regarded an extensive application of the quantum of action "quite risky." A further investigation of possible influences might include the lectures presented by Lorentz in Göttingen (October 24-29, 1910); it was certainly no later than on this occasion that Sommerfeld was exposed to the ideas of the young physicist Haas and incorporated them into his thinking.

In his interview for the Sources for History of Quantum Physics, Sommerfeld's student Paul S. Epstein relates a conversation held with Sommerfeld which, more clearly than anything else, marks the great turning point: "So they [Sommerfeld and Debye] did not see the necessity of the quantum until Sommerfeld—it must have been 1911—needed a rest, though it was in the middle of the year, and for a few days he went to Einstein at Zürich. The idea of recreation was to him to talk the whole day physics with Einstein. And there he told Einstein, that Debye was enchanted with the theory of J. J. Thomson* [Planck's formula without the quantum] . . . so they got that volume [with the paper of Thomson], and then they read it together. And Einstein laid his finger on the error. So J. J. Thomson* was punctured . . ."[15]

On questioning, Epstein admitted that the conversation between Sommerfeld and Einstein in Zürich might have occurred as early as 1910. If we apply to 1910 Epstein's statement that "it was in the middle of the year" (together with the comment that it was unusual that Sommerfeld needed a vacation),† then this visit fits very well into the historical development.

It is even possible that Sommerfeld's visit to Zürich was actually made in connection with Debye's first quantum study (to which Epstein alludes): Debye's paper was received by the *Annalen der Physik* on October 12, 1910; it is plausible that Sommerfeld, be-

*Apparently he meant James H. Jeans and his radiation theory.
†Einstein transferred from Zürich to the University of Prague in March 1911; thus, 1911 could not have been the year of Sommerfeld's visit.

fore releasing the paper for publication, felt the need to discuss it with Einstein in September 1910. He greatly revered and admired Einstein; thus, in a sense, he wanted to obtain the opinion of the "oracle" before finally proceeding with the publication.

A letter written by Einstein to Johann Jacob Laub verifies Epstein's report.[16] It states that Sommerfeld spent an entire week with Einstein "in order to discuss the light question and some matters regarding relativity. His presence was a true joy for me. To a large extent, he agreed with my viewpoint concerning the application of statistics."

3. Debye's Derivation of Planck's Radiation Formula

In 1910, Debye presented possibly the shortest and clearest derivation of Planck's radiation formula. His point of departure was the fact, emphasized particularly by Einstein in 1909, that the two fundamental equations used in Planck's derivation contradicted each other (see p. 54). This concerns on the one hand the classical formula that relates the radiation density u to the average resonator energy U

$$u = \frac{8\pi\nu^2}{c^3} U$$

and on the other hand the quantum-theoretical expression for the average energy of the resonator

$$U = \frac{h\nu}{e^{h\nu/kT} - 1}. \tag{1}$$

Thus Debye wrote in an introduction to his paper submitted to the *Annalen* on October 12, 1910: "Planck's law has been fully verified experimentally, yet its derivation contains a weak point in that the two parts on which the proof of the radiation law is based differ in their fundamental assumptions . . ."[17]

In his interview for the Sources for History of Quantum Physics, Debye offered the following pertinent recollection: "My whole business was, Planck is illogical. On the other hand, it looks as if the whole thing is very good. Can we get rid of the illogical part?" And Debye gave an emphatic affirmation of Uhlenbeck's question: "And that is why you refused to use this connection between radiation intensity and average energy of the oscillators?" Debye: "Yes, yes, yes, yes, yes, yes."[2]

Debye used the Rayleigh-Jeans summation of the resonance vibrations of the radiation cavity. It is well known that this density of modes is $Z = (8\pi/c^3)\nu^2 d\nu$. Debye only had to assign an average energy (1) to each degree of freedom in order to directly obtain Planck's radiation formula.

The letter written by Debye to Sommerfeld on March 2, 1910, already provides a very clear indication of this fundamental idea. At the historically significant point, Debye states: "The answer is as follows[:] Each degree of freedom no longer is assigned the same amount of energy; rather, each degree of freedom is assigned, given its frequency of oscillation ν, the energy

$$\frac{h\nu}{e^{h\nu/kT} - 1}.$$

If this is taken as a starting point, it is easily possible to apply Jeans's viewpoint; it leads directly to Planck's law. The beauty of it is that no intermediate use of resonators is required nor must Planck's averaging procedure be used."

4. Sommerfeld's First Quantum Study

Toward the end of 1909, Sommerfeld had remarked in the *Physikalische Zeitschrift:* "It seems to me that . . . the brems component of radiation has nothing to do with the quantum of action." On January 7, 1911, he commented: "I should like to use this opportunity to withdraw . . . my remark. At that time, I was still under the generally accepted impression that the quantum of action could only be considered for processes occurring at a definite frequency. Now I realize that our brems duration τ can be substituted for Planck's $1/\nu$."[18] In his presentation to the Bayerische Akademie on January 7, 1911, concerning the structure of γ rays, he particularly tried to specify the electromagnetic radiation resulting from electron accelerations by the use of a quantum expression.

The γ radiation of radioactive atoms had been of interest to Sommerfeld for some time; his correspondence with Friedrich Paschen during 1904/1905 dealt in considerable detail with the nature of γ rays. Now, in December 1910, Sommerfeld had calculated a complete theory of β decay which of course, as we know today, was quite premature.

Not even the concept of the atomic nucleus was available to Sommerfeld at that time. In his view, the emitted electron is acceleration, the final velocity (viz., the energy E_β) of the electron final velocity with which it leaves the atom; electromagnetic radiation is generated by this acceleration of the electron. Thus, β and γ emissions are mutually related.

On the one hand this acceleration determines the final velocity of the electron, and on the other hand the electromagnetic radiation. Under certain assumptions concerning the type of acceleration, the final velocity (viz., the energy E_β) of the electron and the γ energy E_γ are therefore also interrelated. Thus, Sommerfeld obtained the formula[19]

$$\frac{E_\beta}{E_\gamma} = \frac{6\pi m_0 c^2 l}{e^2} \cdot \frac{\sqrt{1 - \beta^2}}{\beta}.$$

Here, in the usual manner, β refers to the velocity (measured in units of velocity of light c) with which the electrons leave the atom, m_0 is the rest mass of the electron, e the charge of the electron, and l the path length of acceleration. If the radius of the sphere of action, that is, $l \sim 10^{-8}$ cm, is substituted for the acceleration path length l, then one obtains a value of E_β/E_γ which is too large by a factor of 10^3 to 10^4; expressed another way, an appropriately smaller value than 10^{-8} cm must be chosen for the acceleration path length l.

Sommerfeld now noted that, in order to determine this atomic (or rather subatomic) length, it is possible to apply Planck's quantum of action (compare this with the later considerations of

Bohr, p. 148). Thus Sommerfeld at this point first introduced the quantum concept into his thinking in order to reach a reasonable order of magnitude for E_β/E_γ in his calculations:[19] "While our considerations up to this point have been quite hypothetical, we nevertheless wish to attempt, through the introduction of a new hypothesis, to carry them a step further in order to express the ratio E_β/E_γ by known quantities and entirely as a function of velocity. For we are applying the fundamental hypothesis of Planck's radiation theory to radioactive emissions and assume that a quantum of action h is given off in each such emission. We set the 'action' of an emission (time integral of energy) equal to an acceleration time τ times the emitted total energy . . . ; thus

$$\tau \cdot \frac{m_0 c^2}{\sqrt{1 - \beta^2}} = h."$$ (2)

This formula provided, in the desired manner, a determination of brems (deceleration) time τ, that is, of the brems path length l.

Equation (2) is Sommerfeld's fundamental hypothesis concerning the interaction between electron and atom; thus it should also—as Sommerfeld explained in greater detail late in 1911—apply to bremsstrahlung, the photoelectric effect in metals, and the photoelectric effect in molecules (ionization).

Sommerfeld even hoped to be able to derive Planck's radiation formula by means of his equation. Planck's oscillators, which are in thermodynamic equilibrium with the radiation, must also be able to exchange energy with each other. This interaction should be transmitted in the usual manner by free electrons; thus this interaction should be a "photoelectric effect in the atom". But since Sommerfeld's equation (2) applies to this process, Sommerfeld suspects "that the quantum characteristic of black-body radiation might be due to the possibility that it is accompanied by a continuous photoelectric activity of the atom."[20]

As Sommerfeld noted (on the margin of Planck's letter of April 6, 1911), in his view "only β-ray emission has a quantum nature"; in his letter to Planck dated April 24, 1911 he is more specific: "A molecule always emits and absorbs an electron according to a quantum process $h;$ this completely determines the corresponding electromagnetic radiation emitted."[21]

As previously mentioned, Sommerfeld believed that his fundamental hypothesis applied to all elementary processes; according to him, even the emission and absorption of electromagnetic radiation by the atomic resonators should always occur together with the absorption or emission of electrons. Planck had this to say: "As regards the emission of electromagnetic radiation, I have no objection to the view that this is always caused by [emitted or absorbed] electrons . . . But so far as I can tell, electrons do not play a significant role in the absorption of electromagnetic radiation freely propagating in space; that is, we cannot say that every time that free electromagnetic radiation is absorbed, an electron is emitted."[22]

This shows that the mechanism of fundamental atomic processes was not yet clearly understood around 1911; it was this ignorance which primarily prevented the breakthrough to quantum-theoretical dynamics of the atom (see p. 99).

5. Planck's and Sommerfeld's Hypotheses

Sommerfeld first presented his fundamental quantum-theoretical hypothesis to the Bayerische Akademie in Munich on January 7, 1911; barely a month later, Max Planck also presented a new radiation hypothesis in Berlin at a meeting of the Physikalische Gesellschaft. Since the end of 1909, Planck had again devoted himself intensively to the quantum problem, recommending both to himself and to others that risky steps be avoided as much as possible: "The introduction of the quantum of action h into theory should be carried out as conservatively as possible; that is, only such changes should be made in the existing theory as have been shown to be absolutely essential."[23]

These words of Max Planck indicate the extent to which his thinking was deeply rooted in the classical principles of physics. Aside from his knowledge, which in this case proved to be a heavy burden, his attitude was also stamped by his personality. A distinguished manner as well as a reserved, cautious nature, was fundamental to his character. Planck's new radiation hypothesis corresponded to his conservative disposition. In it, he retracted some of his own quantum views of 1900: Only the emission of electromagnetic radiation by oscillators is still considered a quantum process while absorption takes place continuously.

Planck's argument for continuous absorption agrees completely with the classical tradition in the field of physics: "The absorption of a finite quantum of energy from a finite amount of radiation can only occur in a finite period of time which will be all the greater, the smaller the intensity of the exciting vibration J is in comparison to the energy element ϵ. While the energy element $\epsilon = h\nu$ continues to increase in value with increasing vibrational frequency, the intensity J_ν decreases so rapidly that the above-mentioned time will finally become enormously large for short wavelengths. This actually contradicts the given presupposition; for if the oscillator had begun to absorb energy so that the impinging radiation is interrupted, the oscillator would be absolutely prevented from attaining its full quantum of energy."[24]

In 1914, Planck also withdrew quantum emission: "In addition to the absorption, the emission of thermal radiation is also assumed to be continuous." In line with Sommerfeld's hypothesis, Planck here assumed that "the quantum effect does not occur between the oscillators and the radiation but solely between the oscillators and the free particles (molecules, ions, electrons) which exchange energy with the oscillators during collision."[25]

But this third hypothesis of Planck hardly played any role in the further development of quantum theory. Within the context of our discussion, it will only be necessary to cite Planck's second hypothesis of quantum emission (first stated on February 3, 1911).

After Sommerfeld sent a preprint of his paper on the structure of γ rays, with an accompanying letter, to Planck on March 12,

1911, the latter compared his own hypothesis of quantum emission with Sommerfeld's fundamental hypothesis. He undertook this comparison in two letters addressed to Sommerfeld; in view of their special significance, the two pertinent sections will here be quoted at length:

"But now to the question concerning the relationship between your results and my new hypothesis of quantum emission. For the time being, I see no contradictions; it should only be remembered that you are invariably dealing with a single emission event while my hypothesis refers to a number of emissions. This relates to the fact that you calculate with time spans of the order of magnitude of the 'acceleration time' τ, while my time spans are always large with respect to τ. Whereas you go into details concerning a single emission event, such a single emission from my point of view is an unusual occurrence of negligibly small time duration about which no further details can be gathered from my hypothesis. The latter has only significance when dealing with several emissions, since it states that these emissions take place independently of each other. But this agrees quite well with your considerations. That the absorption of energy is always continuous and does not occur in quanta also appears to be compatible with your calculations. . . .[26]

"For the present, just a few words concerning the positions of our two concepts with respect to each other. Yours is formulated in your letter [of April 24, 1911] as follows: 'A molecule always emits and absorbs an electron according to a quantum process h; this completely determines the corresponding electromagnetic radiation emitted.' My hypothesis, on the other hand, states:

1. The vibrational energy of a molecule vibrating at an eigen frequency ν, need not be a whole-number multiple of $h\nu$.

2. The emission of electromagnetic radiation by such a molecule occurs in whole-number multiples of the energy quantum $h\nu$.

3. The absorption of electromagnetic radiation by such a molecule

is directly proportional to the impinging energy; that is, it does not occur in energy quanta.

"Our two hypotheses cannot be directly compared with each other, because you are talking about an arbitrary number of molecules while I am concerned with molecules vibrating periodically and also because you are considering electrons while I am looking at electromagnetic radiation."[22]

Not only Planck, but Sommerfeld also compared his application of the quantum of action with Planck's method of energy quanta and determined that both concepts are alien to classical electrodynamics; his views, however, unlike those of Planck, were reconcilable with electrodynamics.

"In fact, to explain the origin of x rays, we determined the brems duration from the hypothesis of the quantum of action, whereas we determined the energy and the structure of the x rays corresponding to the brems process from classical electrodynamics. The brems duration is a parameter that enters into the electromagnetic field and without whose knowledge the latter becomes indeterminate. But electrodynamics is not able to determine this parameter in and of itself. . ."[27]

Thus, Sommerfeld believed that his equation (2) supplemented classical electrodynamics exactly at the point where something was needed but otherwise left it completely undisturbed. In his view, such a solution of the quantum problem would also have been the most satisfactory: classical physics and all the "proudest achievements of physics, yes, of all the natural sciences," as Planck put it, could be maintained without change; only a few new floors would have been added to this "very well founded edifice."

In the present author's view, such wishful thinking—in addition to technical considerations—determined Sommerfeld's opinions, just as Planck who, after introducing the quantum hypothesis in 1900, searched for a way to maintain the old theory.

6. Application and Criticism

In a letter of April 6, 1911, Planck wrote to Sommerfeld regarding the latter's equation: "I would have been inclined, instead of writing

$$h = \tau \cdot \frac{m_0 c^2}{\sqrt{1 - \beta^2}} \tag{2}$$

to write the expression:

$$h = \int_0^\tau dt \, m_0 c^2 \sqrt{1 - \beta^2} \tag{3}$$

(β is given as a function of t); at least, this is what I attempted to do in my dynamics of systems in motion [*Annalen der Physik*, vol. 26, p. 23, 1908] but I was not able to come up with anything worthwhile."[26]

At this point in Planck's letter, Sommerfeld made a notation of the word "relativity"; also, he used this equation of Planck in his later publications. Thus, it was Planck from whom he obtained the suggestion for using a relativistically invariant expression. Sommerfeld now intended to carry this "relativity" idea forward in order to obtain a justification for the equation. His attempt was unsuccessful; Sommerfeld gets lost in rather formal abstractions. He finally wrote his equation with H as the "kinetic potential":

$$\int_0^\tau H \, dt = \frac{h}{2\pi}, \tag{4}$$

adding the factor $1/2\pi$, or sometimes $1/4$, on the right-hand side of the equation.

Sommerfeld's own comments were: "For the most part we shall view H as a mere abreviation for $T - V$." Generally, he carried out his calculations in a nonrelativistic manner and also set the potential energy $V \equiv 0$. The following relationship then holds:

$$\int_0^\tau T(t)\, \mathrm{d}t = \frac{h}{2\pi}. \tag{5}$$

Since the details of the acceleration and deceleration processes are not known, a definite course must be postulated for the acceleration. Thus, no additional uncertainty is introduced by the integral equation; the special course of the acceleration is expressed only by a constant factor.

If, for example, a constant acceleration is assumed and if T designates the kinetic energy of the electron prior to the brems process, then from (5) it follows that

$$T \cdot \tau = 3\, h/2\pi. \tag{6}$$

Since the values to be calculated are radiation intensities which at that time could be measured in orders of magnitude only, Sommerfeld can, in place of (6), write the even simpler equation

$$T \cdot \tau = h. \tag{7}$$

Here, Sommerfeld stated with regard to x-ray bremsstrahlung: "High-energy cathode rays are decelerated in a short time span while low-energy cathode rays require a longer time. This result, with which we are quite familiar from experience, is nevertheless quite peculiar. It contradicts every analogy in the field of ballistics . . ."[28]

Sommerfeld used his equation in 1911 to carry out extensive calculations on the photoelectric effect (see p. 126), on the theory of γ rays, on collision ionization, and on x-ray bremsstrahlung. He worked out these calculations in considerable detail; this, as we know today, was a premature effort. We shall take a closer look at Sommerfeld's work at this point, using the example of x-ray bremsstrahlung.

What experimental result was Sommerfeld able to derive from his formula (7)? He answered this question himself later in the second volume of his book *Atombau und Spektrallinien*: "Of course, this postulate still was unable to explain the fundamental equation of the short-wavelength limit of the spectrum. But it provided a simple explanation for two other experimental results: a, the proportionality between the x-ray intensity and the squared value of tube potential, and b, the low efficiency in the conversion of cathode-ray energy into x-ray energy."[29]

Already during the discussion following Sommerfeld's presentation at Brussels, Albert Einstein had shown that the "important result of Herr Sommerfeld," namely the formula for the x-ray energy, can also be derived in another manner: "I wish to point this out in order to prevent that the satisfactory agreement of the theoretical formula with experimental results be viewed as a direct confirmation of the underlying equation [30]

$$\int (T - V)\, dt = h/2\pi."$$

In contrast to Sommerfeld, Einstein assumed the brems process to be a singular event; that is, this process occurs instantaneously. He now developed the step function $v(t)$ according to Fourier, over an arbitrary interval. (This calculation is given in Sommerfeld's *Atombau und Spektrallinien*.[29]) In calculating the total radiated energy, a characteristically divergent expression is formed. Einstein forced the expression to converge by "cutting off" the higher frequencies in the Fourier series: Only those frequencies v should appear in the bremsstrahlung which satisfy the "quantum condition" $hv \leqslant T$. Aside from insignificant numerical factors, Einstein's calculations, which were carried out on an entirely different basis, led to the same expressions for radiated bremsstrahlung energy as had been determined by Sommerfeld.

7. Historical Effects

Sommerfeld often undertook detailed calculations on problems that required further clarification of fundamentals, such as in his papers on electrons moving at speeds faster than light and particularly in his later work on the Bohr-Sommerfeld atomic model. To solve the principal problems of physics was simply not within his capabilities; rather, he was best suited to provide the mathematical development of a theory and its actual solution. Here is Max Born's commentary on Sommerfeld's book *Atombau und Spektrallinien*, written on March 5, 1920: "You present some matters in such a way that the uninformed reader would be led to believe that everything is already in order; but this is often not the case. For example, the molecular model of H_2 etc., as well as all of the theory of x-ray spectra. Landé, for one, recently pointed out to me quite clearly that actually everything is still in disarray. Would it not be a good idea to emphasize the doubtful aspects more strongly?"[31] In a letter written to Albert Einstein on January 11, 1922, Sommerfeld compared his approach to the problems of physics with that of Einstein: "I am only capable of developing quantum techniques; you must provide the philosophical basis."[32]

Thus Sommerfeld was most successful where the physical basis of a problem was certain; on the shaky ground on which the quantum concept still rested in 1911, his extensive calculations were decidely premature. His mastery of mathematical methods may have been a significant factor which led him into making these attempts.

On the other hand, Sommerfeld's historical influence can hardly be overstated. In Brussels, his intercession in favor of the quantum theory was a luminous encouragement to others. Léon Brillouin for example, described the impression which Sommerfeld made on three French theoreticians, Poincaré, Langevin, and Marcel Brillouin, who had participated in the Solvay Congress: "They were all very much impressed with this Sommerfeld discussion . . . and this was one of the reasons I decided to go to Munich."[33]

Sommerfeld's work after 1911 had a strong positive influence on the climate of opinion regarding the quantum concept. He provided a stimulus to many physicists, and they looked to his authority for support of their ideas. Though his equation and his calculations did not agree with reality, they were extremely productive in their historical effect of providing possibilities for many significant applications.

In one important respect, Sommerfeld's interpretation of the elementary quantum of action as a new physical fact incapable of further derivation was well-founded. At the Solvay Congress in Brussels, he remarked:

"But I do not want to go so far as to interpret this relationship between the quantum of action h and molecular size as the actual origin of h, as Haas and probably also Lorentz are inclined to do . . . Aside from the uncertainty of our knowledge of molecular dimensions, it seems to me that such a viewpoint does not do justice to the universal significance of h. I would be inclined to give preference to the opposite point of view: not to explain h on the basis of molecular dimensions but rather to view the existence of the molecule as a function and result of the existence of an elementary quantum of action. An electromagnetic or mechanical 'explanation' of h appears to me quite as unjustified and hopeless as a mechanical 'explanation' of Maxwell's equations . . . . If, as can hardly be doubted, physics needs a new fundamental hypothesis which must be added as a new and strange element to our electromagnetic view of the world, then it seems to me that the hypothesis of the quantum of action fulfills this role better than all others."[34]

References

1. Armin Hermann, *Physikalische Blätter*, vol. 23, 1967, p. 442–449.

2. Interviews with the Sources for the History of Quantum Physics. Manuscripts in the Niels Bohr Archives, Copenhagen (oral records): Peter Debye.

3. Armin Hermann, Die frühe Diskussion zwischen Stark und Sommerfeld . . . In: *Centaurus*, vol. 12, 1967, p. 40.

4. Peter Debye, *Physikalische Zeitschrift*, vol. 9, 1908, p. 773.

5. Ref. 3, p. 51.

6. Johannes Stark, *Physikalische Zeitschrift*, vol. 10, 1909, pp. 579–586 and 902–913.

7. Walther Friedrich, *Annalen der Physik*, vol. 39, 1912, pp. 377–430.

8. Ref. 3, p. 45.

9. Arnold Sommerfeld, *Physikalische Zeitschrift*, vol. 10, 1909, p. 970.

10. Max von Laue, Gesammelte Schriften und Vorträge, vol. 3, Braunschweig 1961, p. XXII.

11. Ref. 9, p. 976.

12. Max Planck, Physikalische Abhandlungen und Vorträge, Braunschweig 1958, vol. 2, p. 242.

13. Arnold Sommerfeld, *Sitzungsberichte der Math.-Phys. Klasse der Königlich Bayerischen Akademie der Wissenschaften*, vol. 41, 1911, pp. 1–60.

14. Peter Debye, letter to Arnold Sommerfeld, Sommerfeld estate, March 2, 1910.

15. Ref. 2, Paul S. Epstein.

16. Albert Einstein, letter to Johann Jacob Laub, Copy at the Library of the Federal Technical Institute (ETH) Zürich, Einstein Collection, September (?) 1910.

17. Peter Debye, *Annalen der Physik*, vol. 33, 1910, p. 1497.

18. Ref. 13, p. 43, footnote.

19. Ibid., p. 24.

20. Die Theorie der Strahlung und der Quanten, Arnold Eucken, ed. [Transactions of the 1st Solvay Congress], Halle 1913, p. 303.

21. Arnold Sommerfeld, letter to Max Planck, missing. Partial reconstruction out of Planck's letters (cf. refs. 22 and 26).

22. Max Planck, letter to Arnold Sommerfeld, Sommerfeld estate, July 29, 1911.

23. Ref. 12, p. 247.

24. Ibid., p. 253.

25. Ibid., p. 330.

26. Max Planck, letter to Arnold Sommerfeld, Sommerfeld estate, April 6, 1911.

27. Ref. 20, p. 294.

28. Arnold Sommerfeld, *Physikalische Zeitschrift*, vol. 12, 1911, p. 1062.

29. Arnold Sommerfeld, Atombau und Spektrallinien, vol. 2, 3rd edition, New York 1951, p. 496.

30. Ref. 20, p. 308.

31. Max Born, letter to Arnold Sommerfeld, Sommerfeld estate, March 5, 1920.

32. Albert Einstein/Arnold Sommerfeld, Briefwechsel, Armin Hermann, ed., Basel 1968, p. 97.

33. Ref. 2, Léon Brillouin.

34. Ref. 20, p. 290.

7

Walther Nernst

1. Experiments on Quantum Phenomena

Albert Einstein had already analyzed the photoelectric effect in 1905 from a quantum standpoint (see p. 62). His light-quantum concept allowed a very simple point of view whose conclusions, however, were in marked contrast to the concepts of classical physics. For didactic reasons such ideas are still used today in presenting an introduction to quantum theory. The reference to "Einstein 1905" is then sometimes misunderstood to mean that this effect, discovered by Heinrich Hertz and refined by Wilhelm Hallwachs and Philipp Lenard, constituted a proof for the quantum theory as early as 1905. Actually, years would elapse before the verification of Einstein's relationship between the frequency of incident ultraviolet radiation ν and the maximum velocity v of the emitted electrons

$$\frac{mv^2}{2} = h\nu - P$$

was experimentally confirmed. As late as 1913, Robert Pohl and Peter Pringsheim stated:

Joffé has . . . determined that the values of initial photoelectric velocities published by Erich Ladenburg in 1907 can be interpreted just as well by Einstein's formula as by the equation originally assumed by Ladenburg himself, according to which the velocity rather than its squared value should be proportional to frequency. Since then, a number of studies have appeared whose intention was a final solution of this problem without, however, going significantly beyond the results of Ladenburg and Joffé as far as we can determine.[1]

A similar opinion was expressed by Joffé early in 1913: At some future time "it might be possible to reach light intensities, particularly of very short wavelengths, which will allow observation of the discontinuity, assuming that it exists. But this goal has not even been approached; pertinent observations are still in the preliminary investigation stage."[2]

This situation is well described by R. A. Millikan in a retrospective view expressed in 1916: "During the course of the last ten years I have been engaged in investigations related to Einstein's equation for the photoelectric effect in the hope of finding its experimental verification. At times, the results seemed to me irreconcilable with that equation . . . "[3]

Not until about 1915–16 was a full experimental verification of the photoelectric effect provided. It was only after the development of Bohr's theory that this effect was used for a precise determination of Planck's quantum of action.

While some doubts existed concerning the validity of Einstein's relation, there was greater agreement among physicists that in any event the maximum energy of photoelectrically emitted electrons did not (significantly) depend on light intensity (as was to be expected on the basis of classical concepts), but that it is the frequency of light that is the factor determining the electron velocity.

This circumstance was explained as a resonance phenomenon, quite in keeping with the classical point of view. This, for example, was reported by Karl Markau to the congress of physical scientists in 1908 at Cologne from experiments which were carried out by him together with Erich Ladenburg. Here is a summary of their experimental results: "The photoelectric effect is purely a resonance phenomenon; the electrons are excited into resonant vibrations by light of some definite frequency that is identical to the electron oscillation frequency. These electrons are then emitted from the irradiated metallic plate at a velocity directly related to the vibration frequency. The greater the latter, the greater the velocity."[4]

The young physicist Max Born likewise interpreted the photoelectric effect as a resonance phenomenon; it is even possible that he is responsible for Markau's explanation. He admitted[5] that he was also guilty of what was later considered the "crime" of viewing the photoelectric relationship between the frequency of light and the energy of electrons in a classical manner. He states that at a meeting of the Physikalische Gesellschaft in Berlin he made a remark to that effect during a discussion following a lecture; apparently this remark made a significant impression on Planck.

In 1911, Sommerfeld in a joint paper with Peter Debye[6] applied his fundamental hypothesis to the photoelectric effect. Here he assumed a position somewhat conciliatory between the "view of electrons released by a resonance effect" and the light-quantum hypothesis.

Even though Sommerfeld used a quantum equation for treating the effect by use of his fundamental hypothesis $T \cdot \tau = h$ (see p. 119) his viewpoint, nevertheless, was still essentially classical. This is particularly shown by his expression[7] for "accumulation time" $\tau = \sqrt{16mvh/eE}$ which increases with decreasing amplitudes E of the exciting electromagnetic wave. This is intended to be the time span between initiation of irradiation and emission of the electron; the time during which energy is "accumulated" in the metallic plate.

When Edgar Meyer and Walther Gerlach undertook their measurements on the photoelectric effect late in 1912 or early in 1913, they based their work primarily on these views of Sommerfeld.* In particular, they considered the measured time delay (between the start of radiation and the detection of emitted electrons) as being due to an energy accumulation.

On March 7, 1913, Meyer and Gerlach reported their measurements at a meeting of the Swiss Physikalische Gesellschaft in Zürich.[8] During the discussion, Einstein commented that an investigation should be made to determine whether some kind of secondary effects might be taking place in the gas as a result of Brownian motion. This is reported by Gerlach himself in an interview for the Sources for History of Quantum Physics:

A few days later, Meyer visited me and told me that Einstein

---

*Although the definitive publication of Debye and Sommerfeld was only sent to the *Annalen der Physik* at the beginning of April of 1913, Sommerfeld's views were known because he had previously presented his fundamental ideas at the first Solvay Congress in Brussels early in November 1911. The proceedings of the Congress were published in mid-1912.

had sent him a little note. This note no longer exists . . . It was written on a piece of toilet paper with a few equations concerning Brownian motion. So we told ourselves that the only possibility to find out about it is to investigate the photoelectric effect on ultramicroscopic particles at low pressures. Then . . . [the time span] became increasingly shorter, since the electrons then moved away without returning as a result of Brownian motion. Of course I thought along quantum lines at that time . . .[9]

Contrary to the view of Debye and Sommerfeld as well as of Pohl and Pringsheim, Gerlach and Meyer were able to point out the secondary nature of the time delay which "cannot be directly due to the photoelectric effect."[10] But that only occurred in October 1913.

What, then, was the status of experimental investigations regarding the other "quantum phenomena"? As Einstein had pointed out in 1905, the ionization of gases by ultraviolet light (the photoelectric effect in molecules) had already been verified by Philipp Lenard. But, as Gerlach remarked as late as 1921, it had by no means been possible "to carry out these extremely difficult measurements in such a way as to permit quantitative conclusions."[11]

Physicists were also convinced that a quantum law applied to the interactions between atoms (or molecules) and electrons (Stark, 1908; Franck and Hertz, 1911), but the systematic electron collision experiments of Franck and Hertz were not initiated until 1911. Definitive experimental results (though their interpretation was not yet completely correct) only became available in early 1914 (see p. 79).

The discovery made by Laue, Friedrich, and Knipping in April 1912, marked the beginning of corresponding experiments in the field of x radiation: the selective reflection of x rays demonstrated by the two Braggs provided for the first time a method for the exact measurement of x-ray wavelengths. In 1915, Duane and Hunt determined the short-wavelength limit of x-ray bremsstrahlung. Nor had the investigations of Emil Warburg on the quantitative validity of the photochemical quantum laws progressed any further.

Today the duality principle is recognized as one of the most significant statements of quantum theory. What was its experimental background? Since 1903, J. J. Thomson had repeatedly pointed out, as a result of certain phenomena (photoelectric effect, ionization of gas molecules) related to the propagation of ultraviolet light, that it was likely that electrical energy is not uniformly distributed in the wavefront but rather occurs in the form of spots: "I think there is evidence that the wave front does in reality much more closely resemble a number of bright specks on a dark ground than a uniformly illuminated area."[12]

Willy Wien and Johannes Stark also called attention in 1909 to the "concentrated energy," particularly in x rays (which was completely incompatible with a decrease of intensity by $1/r^2$). After Laue's discovery of x-ray interferences, which had to be accepted as irrefutable proof of the wave nature of x rays, Rutherford wrote to Bohr on February 24, 1913: "There appears to me no doubt that the x rays must be regarded as a type of wave motion; but I personally cannot escape from the view that the energy must be in a concentrated form."[13] Thus, even the strongly pronounced corpuscular characteristics of x rays were recorded by only a few experimental physicists, without noticeable effect on the development of the quantum concept.

These quantum phenomena, added to the careful experimental investigation of thermal radiation, could therefore offer no proof of the quantum concept during the time period we are considering; at best they provided guidelines to which the various physicists gave more or less weight. As a result, the quantum phenomenon "specific heat" became particularly important. As early as 1907, Einstein had been able to provide the fundamental explanation for the decrease of specific heat at low temperatures (see p. 63). Here, Walther Nernst made a significant contribution, through his extensive measurements, to the development of quantum theory.

2. Measurements of Specific Heat

At the close of the 1880's, Walther Nernst had been successfully engaged in providing a theoretical foundation for earlier experimental chemical knowledge by means of the new "physical chemistry." For a hundred years chemistry had used the concept of selective affinity, and attempts had been made since the middle of the 19th century to identify it with a measurable quantity, namely the heat of transformation of chemical reactions.

After van't Hoff had substituted free energy as the real measure for affinity, Walther Nernst recognized in 1906 that the old relationship *affinity equals heat of transformation*, while not true for any arbitrary temperature, does apply as a boundary condition in the transition to absolute zero.

The significant statement resulting from this is that entropy asymptotically approaches a limiting value independent of pressure, composition, and so on, near absolute zero. Sommerfeld called this "third fundamental law of thermodynamics" the "most ingenious development of classical thermodynamics in our century." In 1906, Nernst drew the further conclusion from his law that the specific heats (of liquids and solids) must, on approaching absolute zero, attain a limiting value independent of the characteristics of the specific material. At that time he did not yet recognize that this limiting value equals zero.

Thus, the behavior of specific heat as the temperature approaches absolute zero was already of great theoretical interest to Nernst as early as 1906. At his institute in Berlin, he undertook, with his associates, the experimentally difficult task of determining the specific heats. Previously, only relatively inaccurate measurements had been carried out—an average value taken over a certain temperature range—and practically no work had been done at extremely low temperatures.

It took three years to complete the preparations. On February 17, 1910, Nernst gave the first report on his results. In the meantime, his attention had been drawn to Einstein's theory of specific heat: "If the data are plotted graphically, then in most cases nearly linear curves are obtained which often show an increasing drop toward low temperatures so that one gets the distinct impression that the specific heats become equal to zero or at least assume very small values at very low temperatures. This is in qualitative agreement with the theory developed by

Herr Einstein; Lindemann and Magnus are busy with the quantitative evaluation of experimental data along these lines . . ."[14]

In March of 1910, Nernst traveled to Zürich in order to discuss
this matter with Einstein, whereupon the latter wrote to his
friend Johann Jacob Laub: "The quantum theory is firmly
established as far as I am concerned. My prediction regarding
specific heats seems to have been brilliantly verified. Nernst, who
has just visited me, and Rubens are vigorously pursuing the experimental verification so that this matter should be clarified
shortly."[15]

Nernst also became convinced of the validity of the quantum
theory at that time. This was to have significant consequences for
its further development. The measurements of the behavior of
specific heats at low temperatures, which had been undertaken in
order to solve quite different problems, were now pushed forward
as rapidly as possible. The initial results showed that the drop-off
in specific heat at very low temperatures predicted by Einstein's
theory actually does occur.[16]

It was fortunate for the further development of quantum theory
that Nernst had undertaken his measurements of specific heat as
early as 1906. Thus, the experiments were far enough advanced
by 1910-11 that it was possible to talk about a certain degree of
experimental verification of quantum theory. In addition to
thermal radiation, there now was a second field of experimental
investigation which could be theoretically explained by—and only
by—the use of the quantum hypothesis.

"One of the glorious pages in the history of these first decades of
the Physikalisch-technische Reichsanstalt will always be," said
Sommerfeld in September 1911, "that it laid one cornerstone of
the quantum theory, the experimental foundation of black-body
radiation. Nernst's institute probably merits equal esteem by
providing us with another, no lesser, foundation stone of quantum
theory through its systematic measurements of specific heats."[17]
Einstein also recognized and appreciated the services rendered by
Nernst; in his words, Nernst has "released all related results from

their theoretical twilight existence."[18] As regards the photoelectric effect, the development of experimental methods for understanding it could not keep up with theoretical developments; here —in the characteristic curve of specific heat at very low temperatures—the experimental results had come to fruition and were somewhat ahead of the theoretical work.

When subsequent measurements of Nernst and his associates showed quantitative deviations from Einstein's formulas, Nernst was not inclined to reject the quantum-theoretical approach of Einstein but rather looked for a new equation on the same basis. Einstein had assumed a single vibrational frequency ν of the solid; by trial and error Nernst and Lindemann found another equation,* considering a second vibration with the frequency $\nu/2$, which satisfied all requirements; it completely reproduced the shape of the specific heat curve.[19]

Apparently both Nernst and Lindemann were convinced of the validity of their equation. "But this will not actually be the case," wrote Einstein to Nernst on June 20, 1911. "Rather, frequencies other than ν and $\nu/2$ will also occur. If, however, your simple scheme is sufficient, we should also be satisfied."[20]

Peter Debye was also troubled by the half quanta of Nernst and Lindemann, as he later recollected.[21] In March 1912, he succeeded—as did at the same time, but independently and in a different manner, Max Born and Theodor von Kármán—in working out a complete and consistent development of Einstein's basic theory.

3. Born and Kármán's and Debye's Theories

Strongly recommended by Sommerfeld for the position vacated by Einstein, Peter Debye on April 1, 1911, assumed the post of professor of theoretical physics at the University of Zürich. His exchange of ideas with Sommerfeld was now continued intensively by correspondence (see p. 103). Debye's lecture at Zürich on

*So that both vibrational energies would have the same magnitude at high temperatures.

thermodynamics provided the occasion for renewed work on the theory of radiation and of specific heat, culminating in the fundamental ideas—directly in line with Einstein's work—for the "Debye theory of specific heat."

On March 9, 1912,* Debye first presented his concepts at the convention of the Swiss Physikalische Gesellschaft in Bern; the first published notice can be found in the *Archives de Genève.*[22] On March 29, Debye also communicated his results to Sommerfeld in a detailed letter:[23]

A short note on my lecture in Bern is expected to appear in the *Archives de Genève.* While I do not yet have proof sheets, I want to outline the highlights to you at this time. Here is how Einstein handled the problem: he takes a body, removes an atom, and assumes that it can be viewed as a resonator (of frequency ν). He then writes for the energy of the individual atom

$$3 \, \frac{h\nu}{e^{h\nu/kT} - 1}$$

His assumption is wrong, as he himself acknowledged later, for there can hardly be any question of a vibrational motion at constant frequency. I therefore proceed in the following manner: I consider the entire body to be like one aggregate molecule. It can vibrate in an infinite number of ways, according to the theory of elasticity. This is actually not true, since it consists of only N atoms and therefore only has $6N$ degrees of freedom. Recognizing this, we must proceed as follows:

1. calculate the vibrational frequencies of the body, taking into consideration its atomic structure, as Jeans did for a hollow cube.

2. assume that each double degree of freedom has the energy

$$\frac{h\nu}{e^{h\nu/kT} - 1} \, .$$

This combination results in the required energy content of the body.

*The paper by Born and Kármán, which evolved at about the same time, was received by the Physikalische Zeitschrift on March 20, 1912. In their later publication they made the following notation: "It seems to us that the priority for an exact formulation and an approximate solution to the problem belongs to Herr Debye, by a few days."

The first item can be solved in approximation by the theory of elasticity. This approximation is adequate for very low temperatures since the highest vibrational frequencies no longer appear there. Just as in the calculations of Jeans, one finds the number of vibration frequencies between v and $v + dv$ proportional to $v^2 dv$ and thus an energy content of about T^4.

The definitive publication of Debye was published a few months later in the *Annalen der Physik*. [24]

Born and Kármán developed their theory of specific heat at almost the same time. In 1911–12, Max Born, lecturer at the University of Göttingen, lived in a house that subsequently also became the home of Theodor von Kármán, an assistant of Ludwig Prandtl at that time. As a result of their conversations at the dinner table, the two became fast friends. Born himself left this report:

We discussed all the urgent problems of physics and mechanics of the day. One topic was Einstein's work on the specific heat of solids. I no longer recall which of us first brought up this subject but I suspect that it was Kármán because he used to follow scientific literature much more thoroughly than I did. [In an interview for the Sources for History of Quantum Physics, Kármán seemed to recall that the idea and initiative came from Born.[25]] We immediately found that Einstein's monochromatic equation had to be improved by taking into consideration the coupling between vibrations in the crystal lattice; here we logically started with the one-dimensional case . . . . We then tackled the three-dimensional problem and found an approximate solution. After we had submitted our study to the *Physikalische Zeitschrift,* Professor Sommerfeld from Munich came to Göttingen to present a lecture; on that occasion we learned that Debye was attacking the problem by using an extremely simple approximation method and had lectured on this subject in Switzerland; his paper was published a short time before ours. Thus, Debye properly deserves the priority by several weeks. Many years later, my associate, M. Blackman, showed by detailed calculations that Debye's approximation, which uses a continuous crystal model and takes into consideration the atomic structure by simply cutting off the vibrational spectrum at a finite boundary, breaks down at low temperatures while the Born-Kármán theory applies in this region as well.[26]

As a result of their studies on specific heat, Born and Kármán had now also become followers of the quantum hypothesis, without really being aware of this themselves: "So far as I recall, we did not experience any great intellectual upheaval but only the joy of having discovered something new ... only later correspondence with Einstein revealed the full meaning of this matter to me; unfortunately, this correspondence* has vanished."[27]

4. The Quantization of Rotational States

Max Planck had introduced the quantum concept in 1900 for a linear oscillator which in later years continued to be the best-known example of a physical system which demonstrates the quantum theory. But this oscillator responded to a definite frequency v only (and did not absorb or emit radiation at other frequencies); thus, it was not possible to identify it directly with real atoms. On the other hand, the treatment of thermal radiation by Planck and that of specific heat by Einstein had shown that it must have many characteristics in common with the atom.

In 1911, Nernst pointed out that not only oscillatory but also rotational movements must be quantized:

We have ... seen that a deviation from the laws of statistical mechanics occurs when we deal with the vibration† of atoms about an equilibrium position. If we generalize the quantum hypothesis to include the absorption of energy in definite quanta not only in vibrations about an equilibrium position but, as appears quite reasonable, to the rotation of a mass, we are led to further conclusions that might help clarify certain contradictions of earlier theory. This would explain, for example, why a molecule of a monoatomic gas is incapable of absorbing noticeable quantities of rotational energy.[28]

After Nernst and Lindemann had attempted to deal with this problem in 1911, Niels Bjerrum took a closer look in 1912 at the various rotational states of a molecule and developed a relationship to their band spectra. In accordance with Nernst's idea, Bjerrum[29] quantized the rotational energy in multiples of

*The available correspondence—consisting of 117 letters of Einstein, Max Born, and Hedwig Born—only starts with 1916. Max Born no longer recalls with certainty whether earlier letters had actually existed.
†The original text refers to "rotations," which is an apparent printing error.

Planck's energy quantum $h\nu$:*

$$(1/2)J(2\pi\nu)^2 = nh\nu,$$

where J is the moment of inertia and n any arbitrary whole number. Thus, Bjerrum obtained the following expression for the rotational frequencies:

$$\nu = nh/2\pi^2 J.$$

"According to this equation the rotational frequencies should vary discontinuously and, in particular, should form a series of differences."[29] The frequencies themselves rather than their differences are identified with the infrared band spectra—a remark which is self-evident when viewed historically. Despite the fact that Bjerrum's equation has nothing to do with reality, it seems important that it describes *different* rotational states of one and the same molecule. For one of the essential assumptions for Bohr's atomic theory was the concept of different atomic states; it is possible that these ideas of Bjerrum had a historically significant effect.

After a period of doing research at Nernst's laboratory, Bjerrum early in 1912 returned to Copenhagen where he held a position as lecturer at the Chemical Institute of the university. It can be assumed that Niels Bohr became familiar with the ideas of his countryman Bjerrum and through him with the thoughts of Nernst (see p. 153).

5. Preparation for the Solvay Conference

For five years after the birth of the quantum theory it remained strictly the private domain of Planck. After 1905 other scientists, starting with Einstein, gradually furnished positive contributions to journals in the field. Pertinent mention of the problem is shown in the correspondence between interested physicists.

*According to the Bohr-Sommerfeld theory, the angular momentum and not the kinetic energy should be quantized; this leads to the rotational frequencies $\nu = n^2 h/8\pi^2 J$. A wave-mechanical correction must be applied to this expression.

All of this remained unnoticed by the general community of physicists. The quantum problem only made its debut on the center stage of physics with Einstein's presentation at a convention of physical scientists at Salzburg on September 21, 1909. By that time, Nernst also took notice of Einstein and his theory of specific heat. His support of the quantum theory, in addition to his experimental and theoretical contributions to the intellectual struggle with the old viewpoints, played a particularly significant role because his manner of organizing a problem of recognized importance complemented the formal and careful reticence of Planck. The fact that the quantum theory gained many new supporters in 1910 and 1911 and moved into the foreground of interest of both physicists and physical chemists can be attributed in no small degree to the services rendered by Nernst. His role[*] is clearly demonstrated in his conception and preparation of an international conference on the quantum problem which involved all the leaders in this field. This meeting became known as the First Solvay Congress and was an important milestone in the development of quantum physics, as Nernst had intended and envisaged.

After Nernst had conceived the idea[†] that an international conference be called together, he prepared a short memorandum which he submitted to Planck, whom he respected as the intellectual leader of this field. Fortunately Planck's reply, in a letter dated June 11, 1910, is still extant. It reads:[20]

Dear Colleague,
Allow me to make a few further remarks in addition to the marginal notes which I took the liberty of adding to your manuscript with your permission.
The broad conception which you have given to your idea is fully worthy of the task whose solution is its object. In this respect

[*]In addition to Planck, Nernst also played an important role in Einstein's appointment to staff membership in the Prussian Akademie der Wissenschaften in Berlin. A postcard of Nernst dated July 31, 1910, indicates that already at that time he contemplated calling Einstein to Berlin.
[†]It is probable that Nernst modeled the conference after the congress of chemists in Karlsruhe of 1860, which led to the final clarification of the concepts of atomic and equivalent weight.

I am in complete agreement with you; however, I cannot conceal my deep concern with regard to the possibility of carrying it out. I have already indicated in my marginal notes that I would expect such a conference to be much more successful if you would wait a while until more supporting evidence is available on this subject.

But in my opinion another point argues even more strongly in favor of postponing such a conference for one year. The fundamental assumption for calling the conference is that the present state of theory, predicated on the radiation laws, specific heat, etc., is full of holes and utterly intolerable for any true theoretician and that this deplorable situation demands a joint effort toward a solution, as you rightly emphasized in your program. Now my experience has shown that this consciousness of an urgent necessity for reform is shared by fewer than half of the participants envisioned by you and that the others would hardly be motivated to attend such a conference. I need hardly mention the older ones (Rayleigh, Van der Waals, Schuster, Seeliger); I doubt that they will ever become enthusiastic about this matter. But even among the younger physicists the urgency and importance of these questions is still not fully recognized. I believe that out of the long list of those you named, only Einstein, Lorentz, Wien, and Larmor are seriously interested in this matter, besides ourselves.

But let another year, or better two years, pass and we will see how the crack which has developed in the theory continues to grow until all those who are now outside the problem will be drawn into it. I doubt whether such processes can be significantly accelerated. The matter must take its normal course. If you will suggest a conference at that time, you will draw the attention of a number of participants a hundred-fold greater and—most important—it will actually take place, something I doubt at this time.

I need hardly add that, regardless of what will be undertaken along these lines, you can be certain of my greatest interest and I assure you that you can count on my full participation in any such undertaking. I can say without exaggeration that nothing has as continuously intrigued and excited me in physics during the course of the past ten years as these quanta of action.

Cordially yours,

Planck.

Nernst interpreted Planck's position as an endorsement. Late in June or early in July 1910, he traveled to Brussels and met the industrialist Ernest Solvay at the home of Robert B. Goldschmidt. Solvay was keenly interested in scientific developments and can be described as amateur scientist. We can safely assume that he must have listened intently to the scientific news reported by Nernst, even though his understanding of the latest developments in the theory must have been cursory. Apparently Nernst was able to describe the revolutionary changes of fundamentals in physics and the necessity for an international meeting so dramatically that Solvay agreed to finance such a conference. On July 26, 1910, he sent Solvay a draft version of the invitation, in German, which Solvay was to sign and transmit to eighteen leading physicists in the field. This draft reads as follows: *[20]

It seems that we are in the midst of revolutionary changes in the fundamentals on which the present kinetic theory of matter is based. On the one hand, a consistent development of this theory leads, unquestioned by anyone up to now, to a radiation equation whose validity contradicts all experimental findings; on the other hand, certain formulas derive from this same theory regarding specific heat (invariance of the specific heat of gases at different temperatures, validity of Dulong and Petit's law down to the lowest temperatures) which are also completely refuted by many measured data.

As shown particularly by Planck and Einstein, these contradictions disappear as soon as certain restrictions are placed on the motion of electrons and atoms in vibration around an equilibrium position (the theory of energy quanta); but this conception is so far removed from the equations of motion used up to now that its acceptance will no doubt involve an extensive reformation of our present fundamental points of view.

In his accompanying letter to Solvay, Nernst suggested that the conference be convened immediately after Easter 1911 and continued: "I would expect and hope that the congress, whose duration should probably be planned for about one week . . . , particularly if the lectures and subsequent debates are to be published in book form, will be a milestone in the history of science."[20] He suggested Brussels as the meeting place.

Solvay agreed to these recommendations and Nernst now undertook the actual preparations. He had originally intended to name Lord Rayleigh as president or chairman of the conference, but then this office was assigned to H. A. Lorentz—a decidedly better

*This is apparently the text that Nernst had already sent to Planck in June and on which the latter had indicated certain changes.

choice. For Lord Rayleigh maintained a very negative attitude toward the quantum concepts and besides, as he himself admitted; he was a "poor linguist" and therefore not effective on occasions of this sort.[20]

Together with Planck and Lorentz, Nernst decided on the agenda and the corresponding speakers and discussion leaders; also, the list of invitations was once more revised. Finally, the appointed lecturers were officially notified on June 9 and the other participants on July 15, 1911. Nernst had expressed the wish that all invitations be sent out by Solvay alone and that his own name not be mentioned. But Nernst's intensive activity could hardly be overlooked by any of the participants who knew very well that he was the actual driving force behind this undertaking—a fact which, no doubt, did not particularly trouble him. On June 20, 1911, Einstein wrote to Nernst: "I gladly accept the invitation to the conference in Brussels and I will be happy to work on the topic assigned to me. I am extremely pleased by the entire undertaking and have no doubt that you are its moving spirit."[20]

The other invited participants also expressed similarly strong approval in their replies. But it is doubtful that all of the enthusiastic responses were due to the subject at hand, for we can hardly assume that they all had harbored a long-standing wish to see such a quantum conference. It is probable that quite a few were moved by a feeling of elation for being included in the small and select circle of participants.

6. The Solvay Conference of 1911
The physicists who met at the Hotel Metropole in Brussels for the opening reception of the Congress on October 29, 1911, were without doubt the leaders in their field. The meeting itself began on October 30 and lasted until November 3.

Only two of the eighteen actual participants,* James H. Jeans

*To this number should be added three editors (R. Goldschmidt, M. de Broglie, and F.A. Lindemann) as well as two guests, both associates of Solvay (E. Herzen and G. Hostelet).

and Henri Poincaré, took a clearly negative position with respect to the quantum concept. Lord Rayleigh, who represented more or less the same point of view as Jeans, did not participate, despite an invitation. A short letter written by him was read at the conference. Poincaré, on the other hand, changed his opinion as a result of the conference.[30] An initially neutral position (because they were unfamiliar or barely familiar with the problems) was taken by five participants: Ernest Rutherford, Marcel Brillouin, Marie Curie, Jean Perrin, and Martin Knudsen. The remaining eleven scientists demonstrated a fundamentally positive position; as could be expected, the viewpoint of the theoretical physicists was much more clearly defined than that of the experimentalists.

Texts of the addresses prepared by each of the twelve speakers had been distributed among the participants some time prior to the conference; the lively and well-informed discussions which invariably followed the lectures showed that at least some of the participants had come thoroughly prepared. A good grasp of the addresses and discussions can still be gained today from the carefully edited transactions, which provide a cross-section of the state of knowledge and opinion existing toward the end of 1911. It is beyond the scope of the present study to go into this material; the interested reader is referred to the easily accessible congress report. The German version is particularly valuable because it includes an appendix in which Arnold Eucken describes the development of quantum theory between the fall of 1911 and the summer of 1913. These transactions provided a wealth of starting points for further investigations of the quantum problem.

As late as the summer of 1910, Max Planck had stated in a letter to Walther Nernst, that "aside from us" only Einstein, Lorentz, Wien, and Larmor were seriously interested in this matter. The present author has doubts about the interest of Larmor, but Johannes Stark and Paul Ehrenfest should probably have been included. But, as we have quoted Planck, "let another year, or better two years, pass, and we will see how the crack which has developed in the theory will continue to grow until all those who are now on the outside will be drawn in to it."

This prediction literally became true. "Drawn into" the struggle for a proper formulation of the quantum concept during the course of 1910 were Arthur Erich Haas and Arthur Schidlof, Ludwig Hopf, Ladislas Natanson, Peter Debye, and Arnold Sommerfeld; while Niels Bjerrum, Frederick A. Lindemann, Friedrich Hasenöhrl, Paul Langevin, Richard Gans; Pierre Weiss, Heike Kamerlingh-Onnes, and many others became involved in 1911. Among the experimental physicists Emil Warburg, James Franck, Edgar Meyer, and Friedrich Paschen should undoubtedly also be mentioned.

The breakthrough was, in fact, achieved in 1911; the climate of thought had become completely transformed. The quantum concept was no longer the view of outsiders but became a matter of significance recognized by many leading scientists. The next generation of physicists were introduced to this problem right from the start; this applies particularly to the students and associates of Sommerfeld and Nernst, both of whom had gathered a large circle of younger assistants around them. The attitude of those under the influence of Sommerfeld was described by Alfred Landé as follows: "The great problem was the quantum-riddle . . . Now at this time [1912–13] in Munich, I remember I was not the only one of this student group who wanted to solve the quantum-riddle . . . . I spent a very great amount of time on this, and . . . there were other students who were also working on it. We almost raced to see who would solve the quantum-riddle first."[31]

The Solvay Congress in Brussels played a crucial role in this development. Physicists who until then had not been involved—such as Henri Poincaré—were won over to the quantum concept. Those already convinced were exposed to an even stronger and deeper impression of the significance of the quantum problem. Thus, Max Planck wrote to Willy Wien on December 8, 1911: "I certainly hope that our scientifically stimulating days in Brussels have also agreed with you . . . the impressions which we had the opportunity of gathering there will give us food for thought for a long time to come."[32] These impressions were subsequently transmitted to their students and colleagues. For example, Sommerfeld lectured on the conference at the Sohncke Colloquium on November 26, 1911. In Manchester, Rutherford gave a report, while in Paris Paul Langevin reported on it in his lecture at the Collège de France. He devoted two sessions to discussing all the papers presented at this Solvay Congress of 1911."[33]

The official report of the congress, which appeared in French during the summer of 1912 and in German translation toward the end of 1913, represented a comprehensive handbook on the quantum problem. Due to the fundamental importance of the questions which it treated and in view of the participation of the leading physicists of the day, its publication strongly stimulated the interest of all physicists involved in new developments. It assured a continuation of the treatment of the quantum problem by a larger number of specialists and favorable conditions for further progress.

These favorable conditions can also be seen from the fact that as a result of the international congress in Brussels the quantum concept now breached the boundaries of the German-speaking world on a broad front and became a subject of interest in France as in England.

It may be of interest to note here that while every university in France offered several professorships in pure mathematics, theoretical physics, by contrast, stood in low esteem as "applied mathematics"; only three professorships existed in the entire country. Physics was essentially in the hands of Paul Langevin, Marcel Brillouin, and Henri Poincaré. Prior to the Solvay Congress only Langevin was a proponent of the quantum concept. The negative attitude of the renowned Poincaré during the Brussels conference was a matter of deep concern for Planck.[34]

In England, James Jeans and Lord Rayleigh had from the outset rejected Planck's derivation of the radiation formula. J. J. Thomson believed that he could explain the stability of the atom on the grounds of classical physics, while Rutherford was too much of an empiricist to recognize the significance of the problem. Bohr noted that "the older Bragg . . . was the only one in England who took any interest in quantum theory."[35]

After the congress we find, for example, that Louis de Broglie "had read and studied in all its details the expositions on quanta to which the first Solvay Congress was devoted." He describes the great impact which the congress report had made on him in the following words: "With youthful vigor, I became enthusiastic about these interesting problems which had been researched, and I promised myself to spare no effort in gaining an understanding of the true nature of these mysterious quanta which Planck had introduced ten years earlier into theoretical physics but whose great significance had not been understood at that time."[36]

For the further development of physics it was extraordinarily significant that toward the end of 1911 and during 1912 the quantum concept made its way into England as well. In June 1912, the British astrophysicist William Nicholson applied the quantum of action to the problems of atomic structure, referring specifically to Sommerfeld (see p. 151).

In a private discussion with Ernest Rutherford, Niels Bohr obtained, toward the end of 1911, "a vivid account" of the discussions held at the Solvay Congress; when the congress report appeared several months later, Bohr studied it closely.[37]

The quantum concept found particularly fertile ground in the intellectual tradition of British science. In England a more phenomenological-empirical orientation obtained and, unlike in Germany, the problem of atomic constitution had already been in the focus of interest for some time.

References

1. Robert Pohl and Peter Pringsheim, Verhandlungen der Deutschen Physikalischen Gesellschaft, vol. 15, 1913, p. 673. A contemporary English translation appeared in *Philosophical Magazine*, vol. 26, 1913, pp. 1017–1024.

2. Abram Fedorovitch Joffé, *Sitzungsberichte der Math.-Phys. Klasse der Königlich Bayerischen Akademie der Wissenschaften,* vol. 43, 1913, p. 31.

3. Robert A. Millikan, *Physikalische Zeitschrift,* vol. 17, 1916, p. 217.

4. Erich Ladenburg and Karl Markau, *Physikalische Zeitschrift,* vol. 9, 1908, p. 828.

5. Max Born, Physik im Wandel meiner Zeit, fourth edition, Braunschweig 1966, p. 224.

6. Peter Debye and Arnold Sommerfeld, *Annalen der Physik,* vol. 41, 1913, pp. 873–930.

7. Die Theorie der Stahlung und der Quanten, Arnold Eucken, ed. [Transactions of the 1st Solvay-Congress], Halle 1913, p. 280.

8. Edgar Meyer and Walther Gerlach, Archives des Sciences Physiques et Naturelles (Archives de Genève), vol. 35, 1913, pp. 398–400.

9. Interviews with the Sources for the History of Quantum Physics. Manuscripts in the Niels Bohr Archives, Copenhagen (oral records): Walther Gerlach.

10. Walther Gerlach and Edgar Meyer, *Verhandlungen der Deutschen Physikalischen Gesellschaft,* vol. 15, 1913, p. 1038.

11. Walther Gerlach, Die experimentallen Grundlagen der Quantentheorie, Braunschweig 1921, p. 104.

12. Joseph John Thomson, On the Light Thrown by Recent Investigations on Electricity on the Relation between Matter and Aether, Manchester 1908, p. 16.

13. Ernest Rutherford, letter to Niels Bohr, Niels Bohr Archives, Copenhagen, February 24, 1913.

14. Walther Nernst, *Sitzungsberichte der Königlich Preussischen Akademie der Wissenschaften*, vol. 1910, p. 276.

15. Carl Seelig, Albert Einstein. Leben und Werk eines Genies unserer Zeit, Zürich 1960, p. 197.

16. Ref. 7, p. 212.

17. Arnold Sommerfeld, *Physikalische Zeitschrift,* vol. 12, 1911, p. 1060.

18. Ref. 7, p. 337 footnote.

19. Ref. 7, p. 215.

20. Jean Pelseneer, Le Premier Conseil de Physique. Unpublished manuscript. Sources for History of Quantum Physics.

21. Ref. 9, Peter Debye.

22. Peter Debye, Archives des Sciences Physiques et Naturelles (Archives de Genève), vol. 33, 1912, pp. 256–258.

23. Peter Debye, letter to Arnold Sommerfeld, Sommerfeld estate, March 29, 1912.

24. Peter Debye, The Collected Papers of Peter J. W. Debye, New York 1954, pp. 650–696.

25. Ref. 9, Theodore von Kármán.

26. Ref. 5, p. 279.

27. Ref. 5, p. 232.

28. Walther Nernst, *Zeitschrift für Elektrochemie,* vol. 17, 1911, p. 270.

29. Niels Bjerrum, Selected Papers, Copenhagen 1949, pp. 34–40.

30. Ref. 7, p. 367.

31. Ref. 9, Alfred Landé.

32. Max Planck, letter to Willy Wien, Staatsbibliothek Preussischer Kulturbesitz, Sign. Autogr. I/285, December 8, 1911.

33. Ref. 9, Léon Brillouin.

34. Max Planck, Manual accompanying the voice recording 15/4 "Stimme der Wissenschaft," Frankfurt about 1958, p. 17.

35. Ref. 9, Niels Bohr.

36. Louis de Broglie, Physicien et Penseur, Paris 1953, p. 458.

37. Niels Bohr, The Solvay Meetings and the Development of Quantum Physics, in: Essays 1958–1962 on Atomic Physics and Human Knowledge, New York 1963, p. 83.

8

Niels Bohr   The Quantum Theory of the
                                  Atom
                                  (1912–1913)

1. The Introductory Work of 1912

While England lagged far behind Germany in the development of the quantum concept, the question of atomic structure had been given much more attention there. Rutherford and his associates at Manchester played a leading role in experimental research.

Here also the many new discoveries made in the rapidly developing field of radioactivity had not yet been formed into a unified pattern. But since the discovery of the wide-angle scattering of α particles toward the end of 1910, Rutherford and his group had become convinced of the planetary model of the atom. Bohr, who worked in Rutherford's institute in Manchester from mid-March until the end of July 1912, also gained this conviction. His first essential contribution, the significance of which can hardly be overemphasized, was that he learned to differentiate between specifically "atomic" and "nuclear" phenomena of atomic physics.

Here are the comments of Georg von Hevesy, in the original English, concerning this discovery of Bohr: "This comes from Bohr no doubt . . . In '12 he had certainly the idea very clear . . . It was a Sunday afternoon. I was at the house of Rutherford. Bohr was also present. I asked Rutherford: 'Alpha particles clearly come from nucleus, no doubt. But where do the beta particles come from?' Rutherford answered: 'Ask Bohr.' Bohr was present and with no difficulty answered that electrons involved in the transmutation process come from the nucleus, and all of the other electrons come from the exterior of the atom."[1]

Those recollections are corroborated by a letter from Hevesy to Bohr dated January 15, 1913: "The question as to where the particles [β radiation] come from, on which you had been working intensely, continues to be of great interest to me. My results tend to show that they originate from within the atom and it would interest me greatly to have your conclusions on this matter. . ."[2]

If, like Bohr, one accepts the planetary model of the atom, then the question of stability becomes acute. The following excerpt from a letter of Hevesy to Bohr[3] dated January 29, 1913 is typical of the views of the (very few) physicists who had become converted to the Rutherford model:

Recently I have also devoted considerable time to the study of Rutherford's atomic model and in particular to the stability requirements. In so doing, I consider the heavy atoms. Light atoms such as hydrogen, for example, lead to great difficulties if, like you, one assumes the hydrogen atom to contain only one electron. How can a positive atomic nucleus be in equilibrium with only a single rotating electron? I simply cannot understand this. What prevents this electron from falling into the center of the atom? One would have to conclude that a light, single, neutral atom cannot exist by itself. For heavy atoms, on the other hand, it is easy to visualize that the repulsion of individual electrons of the ring compensate for the attraction of the nucleus. I would be most interested to learn what your thoughts are in this regard . . .

In contrast to Hevesy, Bohr had already become convinced by mid-1912 that the quantum hypothesis, in particular, could provide the stability required by Rutherford's atomic model. As early as 1911, Bohr's studies on electrons in metals had showed him that a radical revision of classical concepts would be necessary in order to describe atomic phenomena.[4] In 1912 he wrote [verbatim]:

This hypothesis is: that there for any stable ring (any ring occurring in the natural atoms) will be a definite ratio between the kinetic energy of an electron in the ring and the time of rotation. This hypothesis, for which there will be given no attempt of a mechanical foundation (as it seems hopeless), is chosen as the only one which seems to offer a possibility of an explanation of the whole group of experimental results, which gather about and seems to confirm conceptions of the mechanismus of the radiation as the ones proposed by Planck and Einstein.[5]

The preceding comments are in a manuscript which Bohr prepared in June/July 1912 in order to summarize his views to Rutherford. His starting point is the planetary model of the atom, whose essential characteristic Bohr recognized to be that the size of the atom remains indeterminate. What is meant by this is shown more clearly by a letter of Bohr to Hevesy of February 7, 1913: "Contrary to atom-models such as J. J. Thomson's, there is nothing in the quantities determining an atom-model [*sic*] as Rutherford's from which we by help of the ordinary mechanics can determine a length of the same order of magnitude as the dimensions of the atoms."[6] It is possible that Bohr derived this idea from Joseph Larmor, who had already pointed out in 1900 that in an electrically structured atom there exists no quantity of the dimensions of length that could determine the size of the atom.[7]

A relationship just like that discussed by Bohr is given by the formula derived by Haas in February 1910:*

$$h = 2\pi e \sqrt{a \cdot m}$$

Like Haas, Bohr also talked about a definite ratio K between the kinetic energy of an atomic electron and its oscillation frequency:

$$E_{kin} = K \cdot \nu$$

This equation resembles the one given by Haas in February 1910; the latter has the more precise Planck's constant in place of K. Is it conceivable that Haas's ideas contributed to the development of Bohr's theory?

Bohr emphasized on various occasions that he had no knowledge of Haas's work prior to and during the formation of his essential ideas. We have no reason to doubt these statements. Since Haas's paper in the *Jahrbuch der Radioaktivität und Elektronik*[8] was cited in Bohr's first publication of July, 1913, it is evident that Bohr only became aware of this while preparing his paper for publication. This is good evidence that Bohr still followed the literature carefully in March, while editing his work, even after his ideas had crystallized in February 1913. While this would lead us to exclude any direct influence by Haas, an indirect effect

---

*Here, a is the radius of the electron orbit, which Haas takes to be the same as that of the positively charged atomic sphere.

through Sommerfeld is all the more certain. In September 1911, in his address to the meeting of physical scientists in Karlsruhe[9] and again early in November at the Solvay Congress in Brussels,[10] Sommerfeld had emphatically stated his view that the existence of the molecule must be viewed as a function and result of the existence of an elementary quantum of action. Sommerfeld's discussions in Brussels had been particularly well received in non-German-speaking countries.

Early in 1912, William Nicholson referred directly to Sommerfeld: "If, therefore, the constant h of Planck has, as Sommerfeld has suggested, an atomic significance . . ."[11] While Bohr did not mention Sommerfeld's name in his publication of July, 1913, he did refer to the Solvay Congress in Brussels as follows:[12]

The result of the discussion of these questions seems to be a general acknowledgment of the inadequacy of the classical electrodynamics in describing the behaviour of systems of atomic size. Whatever the alteration in the laws of motion of the electrons may be, it seems necessary to introduce in the laws in question a quantity foreign to the classical electrodynamics, i.e., Planck's constant, or as it often is called, the elementary quantum of action.

In an interview for the Sources for the History of Quantum Physics concerning the year 1912, Niels Bohr had this to say: "I was quite clear that the whole thing [the atom] was in some way regulated by the quantum. And I felt that one does see that from Whiddington's experiments*. . . I took the view that we had now the Rutherford atom . . . and that it was regulated by the quantum. Haas had written all these things the year before, but I did not know of them. . . ."[13]

By using the concept that Planck's constant would provide the stability for the planetary model of the atom, Niels Bohr could now turn his attention to the analysis of the simplest possible atom, the atom with a single electron.

Another important discovery of Bohr was that such atoms

---

*The emission of x rays can be excited by cathode rays only if certain minimum velocities are exceeded, in order to excite the corresponding lines.

actually do exist* and that they constitute the element hydrogen. The concept of atomic number of elements—the number of electrons in the atomic shells—was really the essential premise for a successful treatment of the problem of atomic structure, a view which Bohr formed independently. How uncommon the concept of atomic number was at the time is illustrated, for example, by discussions held at the Second Solvay Congress held on October 27–31, 1913, in Brussels.[14]

Thus, toward the end of 1912 Bohr had a clear idea of Rutherford's model and the forces acting within the atom. On November 4, 1912, he reported in a letter to Rutherford: "I have made some progress with regard to the question of dispersion. The number of electrons in a hydrogen- and a helium-atom calculated from the dispersion seems thus to work out nearer to respectively 1 and 2, if the forces acting on the electrons are assumed to vary inversely as the square of the distance, than if they, as in Drude's theory, are assumed to be of the elastic type."[15]

During the course of the year 1912 Bohr had created the basis for successfully dealing with atomic structure. Bohr had become clear on the following points:

1a. The validity of the planetary model of the atom.

1b. The assignment of atomic models with one, two . . . electrons to actually existing atoms.

1c. The separation of "nuclear" from "atomic" phenomena as such.

2a. The ability of Planck's constant to provide stability of the atom.

2b. The needlessness of further explaining such a quantum concept.

Georg von Hevesy lacked insights 2a and 2b and consequently considered complicated atomic systems in which he assumed

*William Nicholson, for example, held that such a system was unstable and discarded it from the very outset.

that a dynamic equilibrium could be established as a result of the interaction of many electrons. Nicholson did not possess the knowledge of point 1b and was not clear with regard to 2a;* he considered models with two, three, four ... electrons which, however, he did not identify as real atoms but rather as a kind of prototype of such atoms.

But Bohr was able to deal with the planetary system having a single electron and to identify it as the model of the hydrogen atom. As indicated by a letter written by him to Rutherford on January 31, 1913,[16] it was the latter who emphatically advised the young Bohr to confine himself to the simplest atomic model and to avoid detailed calculations on complex systems.†

2. The Atomic Energy Levels

Toward the end of 1912 Bohr came across the studies of William Nicholson[17] which must have excited him considerably. "The theory of Nicholson gives apparently results which are in striking disagreement with those I have obtained; and I therefore thought at first that the one or the other necessarily was altogether wrong."[16]

In comparing Nicholson's model of the atom with his own conception, Bohr came to the conclusion that they might be reconcilable. Such a view is only possible by using the concept of various *states* of one and the same atom. In a postal card written on December 23, 1912, to his brother Harald, this fundamental idea is expressed as follows: "Although it does not belong

---

*Nicholson based his analysis on Larmor's condition for the motion of electrons in the atom without emission of radiation. This required that the vector *sum* of the accelerations of the electrons must equal zero in order that no electromagnetic energy be radiated. This condition is not as strong as the condition for radiation required by classical electromagnetic theory but is not adequate for an understanding of motion without radiation. The Larmor condition inhibited Nicholson from considering the simplest planetary system with a single electron.

†The letter referred to includes this statement: "I have tried to deduce some general properties of the systems in question, without—according to your advice—going into detailed calculations on any special system apart from the most simple."

on a Christmas card, one of us would like to say that he thinks Nicholson's theory is not incompatible with his own. In fact, his [Bohr's] calculations would be valid for the final, chemical state of the atoms, whereas Nicholson would deal with the atoms sending out radiation. . . ."[18]

A month later, Bohr developed this view further in a letter addressed to Rutherford on January 31, 1913:

It seems therefore to me to be a reasonable hypothesis, to assume that the state of the systems considered in my calculations is to be identified with that of the atoms in their permanent (natural) state. (This hypothesis seems to be justified by the agreement between the theory and experiments on atom-volumes and Röntgen rays, which I obtained from the first moment and which I have tried to trace still longer.) According to the hypothesis in question, the states of the systems considered by Nicholson are, contrary, of a less stable character; they are states passed during the formation of the atoms, and are the states in which the energy corresponding to the lines in the spectrum characteristic for the element in question is radiated out [sic] . From this point of view, systems of a state as that considered by Nicholson are only present in sensible amount in places in which atoms are continually broken up and formed again; i.e., in places such as excited vacuum tubes or stellar nebulae.[16]

In Bohr's letter to Hevesy of February 7, 1913, the concept of successive binding of the electrons to the atomic nucleus is further clarified. Here, Bohr assumed "that the systems considered are formed by a successive binding of the electrons by the nucleus until the whole system is neutral (compare the formation of a helium atom from an α particle)."

Thus, Bohr had formed the hypothesis in December 1912 and January 1913 that there exist various states of an atom. How did Bohr arrive at these significant ideas?

In the literature we find various starting points which might have been useful, but we have no exact information on what has actually taken place. For example, Willy Wien in 1909 made the following comparison between Planck's oscillators and real atoms in his encyclopedia article:

"Here we face the difficulty that Planck's resonators, like any electromagnetic center of radiation, must respond to electromagnetic oscillations if they can be excited at all. But Planck's theory requires that only a relatively small number be excited, each of which absorbs one energy element while the others in no way respond to the electromagnetic oscillations. Yet this difficulty does not only occur in the theory of radiation; rather it can also be found in the fact that x rays or ultraviolet light fail to ionize all gas molecules, whereas complete ionization would occur if all molecules demonstrated the same behavior with respect to the electromagnetic waves while, in fact, here again only a relatively small number are ionized. There can hardly be any doubt that the two cases are related. This clearly shows that all atoms cannot be considered to be completely identical. Since on the other hand a really permanent difference cannot be assumed, the only alternative seems to be that atoms be viewed as dynamic structures in which the phases of the dynamic processes can differ."[19]

Here, as early as 1909, a parallel was drawn between the various states of Planck's oscillators and the various states of real atoms. In 1911 and 1912, Hasenöhrl and Bjerrum tried to apply a quantum-mechanical treatment to other elementary structures—again, of course, starting out from the well-known oscillator. In this connection it is significant that they were primarily interested in determining the various energy levels.

The phenomena of radioactivity—which at that time must have seemed more interesting to most physicists than the quantum question—also favored a concept of internal degrees of freedom of the atoms. A corresponding hypothesis was formulated by André Debierne* in order to explain why some atoms of a chemically uniform radioactive substance decay immediately while others continue to exist for long periods of time.

While Bohr differentiated between nuclear and atomic phenomena, such concepts would suggest that states of differing energy could apply not only to the atomic nucleus but to the atomic

*During the Second Solvay Congress in Brussels, Marie Curie still adhered to Debierne's view.

shell as well. This thought took hold of Bohr's imagination in the course of his study of Nicholson's work, as evidenced by his letter to Rutherford of January 31, 1913.[16] Since Bohr could not simply assume Nicholson's views on atomic structure (which differed so fundamentally from his own) to be erroneous, his ideas were saved, as it were, by the concept of various atomic states.

As an astrophysicist, it was Nicholson's stated goal to study particularly high-energy forms of matter, such as occur in the sun and the stars.* Bohr therefore applied his ideas directly to the permanent (natural) state of the atom, a state in which the atom does not radiate. This is clearly shown by another section of Bohr's letter to Rutherford of January 31, 1913, cited previously:[16]

I must however remark that the considerations sketched here play no essential part of the investigation in my paper. I do not at all deal with the question of calculation of frequencies corresponding to the lines in the visible spectrum. I have only tried, on the basis of the simple hypothesis, which I used from the beginning, to discuss the constitution of the atoms and molecules in their "permanent" state; it means, I have tried to deduce some general properties of the systems in question without—according to your advice—going into detailed calculations of any special system apart from the most simple.

Thus, on that date Bohr emphasized to Rutherford that he was not involved in the calculation of the line spectra, but five weeks later, on March 6, 1913, he sent Rutherford the first part of his famous "trilogy" containing the theory of the hydrogen spectrum:[20]

As you will see, the first chapter is mainly dealing with the problem of emission of line spectra, considered from the point of view sketched in my former letter to you. I have tried to show that from such a point of view it seems possible to give a simple interpretation of the law of the spectrum of hydrogen, and that

*Nicholson believed that the simple primary atoms (Rutherford's atoms) occur in the plasma (to use a modern expression). Aggregates of such atoms would then form all the atoms existing on earth. Thus, he was concerned with developing a theory of atoms of the earth and actually made such an attempt.

the calculation affords a close quantitative agreement with experiments. (I have given reasons which show, that if the foundation of the theory is sound, we may assume that

$$\frac{2\pi^2 m e^4}{h^3} = 3.290 \cdot 10^{15}$$

Putting your value $e = 4.65 \cdot 10^{-10}$, I get $h = 6.26 \cdot 10^{-27}$

Putting Millikan's value $e = 4.87 \cdot 10^{-10}$, I get $h = 6.76 \cdot 10^{-27}$.

Unfortunately, however, Planck's constant is hardly known with any great accuracy.)

A decisive event must have occurred during February 1913. How was Niels Bohr led to attempt a theory of spectral lines? On October 31, 1962, only three weeks before his death, Bohr was questioned on this subject by Léon Rosenfeld and Thomas S. Kuhn:[13]

Rosenfeld: "How did you come to examine the spectra?"

Bohr: "The spectra was a very difficult problem. There were two different schools—those in England, and then there was the school of the spectroscopists . . . And I discovered it, you see. Other people knew about it, but I discovered it for myself. And I found hydrogen spectrum. I was just reading the book of Stark, and at that moment I felt now we'll just see how the spectrum comes."

Kuhn: "Was this at Manchester that you were reading Stark?"

Bohr: "No, no, that was later in Copenhagen—that was half a year after. It was in January,* I think, of '13 . . ."

The present author must admit that he was very astonished by this information. While for years he has had in his possession the major portion of the scientific papers left by Johannes Stark and has repeatedly pointed out the latter's importance in the early development of quantum theory, he would never have dared suspect such a *direct* influence.

*It is clear from Bohr's letter to Rutherford of January 31, 1913, that this occurred in February 1913.

3. The Reshaping of Stark's Atomic Theory

In 1908 Stark had been the first to develop a clearer picture of the origin of spectra. According to him, a single valence electron is responsible for all lines of a spectrum* which are radiated during the successive transition of the electron from a state of (almost) complete separation to a state of minimum potential energy. Stark again summarized this view in his book *Prinzipien der Atomdynamik II. Die Elementare Strahlung* published in 1911; it was this book that came to Bohr's attention in February, 1913.

The application of Stark's views[21] to the case of the planetary model of the atom which Bohr had already fully accepted for quite some time seems, to the present author, to lead almost inevitably to the following:

1. Since the force with which the electron is attracted to the nucleus is, in Bohr's view, a coulomb force, the electron actually describes an elliptical orbit.
2. After the emission, the orbital path is again elliptical but the ellipse is drawn closer to the nucleus.
3. It can be left open at this point what happens to the electronic path (continuous or in steps) during the emission processes.

Stark's conception, if consistently applied on the basis of Rutherford's model of the atom, necessarily results in a family of ellipses ordered according to the corresponding electron energy.

4. Derivation of the Hydrogen Spectrum

"As soon as I saw Balmer's formula, the whole thing was immediately clear to me," Niels Bohr remarked on several occasions to Léon Rosenfeld.[22] The latter reports a further conversation with Bohr on June 23, 1954, concerning the historical development:[23] "According to Bohr's recollection, he was asked by the

---

*In Stark's view, he was dealing with the band spectrum. But since the mechanism developed by him was only of historical significance (rather than the definite identification of lines), this is a matter of little importance.

young Danish physicist Hans Marius Hansen, who was working as Riecke's assistant in Göttingen in 1911-12, how his atomic theory could explain the spectra. In Bohr's view, these spectra were too complicated. Hansen disputed this and pointed to Balmer's formula."

The present author found among Bohr's books his personal copy of Stark's *Prinzipien der Atomdynamik II*. This copy contained a slip of paper on which Bohr had noted two literature references of papers dating from 1912. Thus it would seem that Bohr obtained these references in connection with his reading of Stark's book. One of the two studies[24] treats the theory of spectral lines and includes the formula

$$\nu = N\left(\frac{1}{n^2} - \frac{1}{m^2}\right).$$

The other reference noted by Bohr is Ludwig Föppl's Göttingen dissertation.[25]

"As soon as I saw Balmer's formula, the whole thing was immediately clear to me": Together with the concept of various elliptical orbits, this formula now provided the transitions between any two elliptical orbits m and n. Since the concept of classifying the elliptical orbits according to energy was apparently already known, Bohr could probably immediately derive the energy emitted by one spectral line

$$h\nu = h \cdot N \left(\frac{1}{n^2} - \frac{1}{m^2}\right).$$

The decisive task still remaining was to classify the various elliptical orbits quantatively according to energy; that is, in order to complete the work, Bohr only had to derive the equation

$$E_n = h \cdot N \cdot \frac{1}{n^2}.$$

As Bohr wrote in his definitive publication *On the Constitution of Atoms and Molecules*, which was completed on April 6, 1913, "the frequency of revolution ω and the major axis of the orbit 2a will depend on the amount of energy W which must be transferred to the system in order to remove the electron to an infinitely great distance apart from the nucleus."[26] Bohr now calculated* ω and 2a as the function of W

$$\omega = \frac{\sqrt{2}}{\pi} \cdot \frac{W^{3/2}}{e \cdot E \sqrt{m}}, \quad 2a = \frac{eE}{W}, \tag{1}$$

where e is the charge of the electron, m its mass, and E the charge of the atomic nucleus. "We see," writes Bohr, "that if the value of W is not given, there will be no values of ω and a that are characteristic for the system in question."[26]

As already indicated in the manuscript in mid-1912 and the letter of February 7, 1913, to Georg von Hevesy, the energy of the atom is determined by a quantum condition.[27] "Putting

$$W = n \cdot h \cdot \frac{\omega}{2}, \tag{2}$$

we get, by formula (1),

$$W = \frac{2\pi^2 m e^2 E^2}{n^2 h^2} \text{ "}.$$

With this, Bohr has fitted the last link into the chain.

This formula—chronologically the last to be determined—opened Bohr's presentation in its first publication in the *Philosophical Magazine*. Starting from the quantitatively determined energy levels of the atom, he arrived at the following result, which later became famous:

"The amount of energy emitted by the passing of the system from a state corresponding to $n = n_1$ to one corresponding to $n = n_2$, is consequently

$$W_{n_1} - W_{n_2} = \frac{2\pi^2 m e^4}{h^2} \left(\frac{1}{n_2^2} - \frac{1}{n_1^2} \right).$$

*For Bohr, ω is the frequency of revolution, not the angular velocity.

If now we suppose that the radiation in question is homogeneous, and that the amount of energy emitted is equal to $h\nu$, where ν is the frequency of the radiation, we get

$$W_{n_2} - W_{n_1} = h \cdot \nu,$$

and from this

$$\nu = \frac{2\pi^2 m e^4}{h^3}\left(\frac{1}{n_2^2} - \frac{1}{n_1^2}\right).$$

We see that this expression accounts for the law connecting the lines in the spectrum of hydrogen."[28]

References

1. Interviews with the Sources for the History of Quantum Physics. Manuscript in the Niels Bohr Archives, Copenhagen, (oral records): Georg von Hevesy.

2. Georg Hevesy, letter to Niels Bohr, Niels Bohr Archives, Copenhagen, January 15, 1913.

3. Ibid., letter, January 29, 1913.

4. Niels Bohr, On the Constitution of Atoms and Molecules, Copenhagen and New York 1963, Léon Rosenfeld, ed., p. XV.

5. Ibid., p. XXIII.

6. Ibid., p. XXXII.

7. Joseph Larmor, Aether and Matter, Cambridge 1900, p. 190.

8. Arthur Erich Haas, Der erste Quantenansatz für das Atom (*Dokumente der Naturwissenschaft*, vol. 10), Stuttgart 1965, pp. 53–60 (reprint).

9. Arnold Sommerfeld, *Physikalische Zeitschrift*, vol. 12, 1911, pp. 1057–1069.

10. Arnold Sommerfeld, in: La Théorie du Rayonnement et les Quanta. Ed. P. Langevin and M. de Broglie. Paris 1912, pp. 313–372. German version: Die Theorie der Strahlung und der Quanten. Ed. A. Eucken. Halle 1913, pp. 252-297.

11. William Nicholson, Monthly Notices of the Royal Astronomical Society, vol. 72, 1912, p. 681.

12. Niels Bohr, On the Constitution of Atoms and Molecules. In: *Philosophical Magazine*, vol. 26, 1913, p. 2. Reprint Ref. 4, p. 2.

13. Ref. 1, Niels Bohr.

14. La Structure de la Matière. Rapports et Discussions du Conseil tenu à Bruxelles du 27 au 31 Octobre 1913. Paris 1921.

15. Niels Bohr, letter to Ernest Rutherford, Niels Bohr Archives, Copenhagen, November 4, 1912.

16. Ref. 4, p. XXXVII.

17. Russell McCormmach, Archive for History of Exact Sciences, vol. 3, 1966, pp. 160–184.

18. Ref. 4, p. XXXVI.

19. Willy Wien, Encyklopädie der mathematischen Wissenschaften, vol. V, 3, Leipzig 1909-1926, p. 355.

20. Ref. 4, p. XXXVIII.

21. Johannes Stark, Prinzipien der Atomdynamik, vol. II. Leipzig 1911, p. 112.

22. Ref. 4, p. XXXIX.

23. Ibid., p. XL.

24. Harold Albert Wilson, *Philosophical Magazine*, vol. 23, 1912, p. 662.

25. Ludwig Föppl, *Crelles Journal für die reine und angewandte Mathematik*, vol. 141, 1912, pp. 251–302.

26. Ref. 4, p. 3.

27. Ref. 4, p. 5.

28. Ref. 4, p. 8f. Bohr's discovery has recently been treated in detail by John L. Heilbron and Thomas S. Kuhn in *Historical Studies in the Physical Sciences* (see p. 161).

Further Reading

Like all historical studies, the present one is based not only on original sources but also leans on numerous other publications, a selection of which is listed below. Some recent items have been included.

John L. Heilbron and Thomas S. Kuhn, The Genesis of the Bohr Atom, in: *Historical Studies in the Physical Sciences*, vol. 1, 1969, pp. 211-290.

Tetu Hirosige and Sigeko Nisio, Formation of Bohr's Theory of Atomic Constitution, in: *Japanese Studies in the History of Science*, No. 3, 1964, pp. 6-28.

Friedrich Hund, Geschichte der Quantentheorie, Mannheim 1967.

Max Jammer, The Conceptual Development of Quantum Mechanics, New York 1966.

Hans Kangro, Vorgeschichte des Planckschen Strahlungsgesetzes, Wiesbaden 1970.

Martin J. Klein, Max Planck and the Beginnings of Quantum Theory, in: *Archive for History of Exact Sciences*, vol. 1, 1962, pp. 459-479.

Martin J. Klein, Planck, Entropy and Quanta, 1901-1906, in: *The Natural Philosopher*, vol. 1, 1963, pp. 81-108.

Martin J. Klein, Einstein's First Paper on Quanta, in: *The Natural Philosopher*, vol. 2, 1963, pp. 57-86.

Martin J. Klein, Einstein and Wave Particle Duality, in: *The Natural Philosopher*, vol. 3, 1964, pp. 3-49.

Martin J. Klein, Einstein, Specific Heats, and the Early Quantum Theory, in: *Science*, vol. 148, 1965, pp. 173-180.

Martin J. Klein, Paul Ehrenfest, vol. 1, The Making of a Theoretical Physicist, New York 1970.

Thomas S. Kuhn, John L. Heilbron, Paul Forman, and Lini Allen, Sources for History of Quantum Physics. An Inventory and Report. Philadelphia 1967.

Russell McCormmach, The Atomic Theory of William Nicholson, In: *Archive for History of Exact Science*, vol. 3, 1966, pp. 160-184.

Fritz Reiche, Die Quantentheorie. Ihr Ursprung und ihre Entwicklung, Berlin 1921.

Léon Rosenfeld, La première phase de l'évolution de la Théorie des Quanta, in: *Osiris*, vol. 2, 1936, pp. 149-196.

Léon Rosenfeld, Max Planck et la définition statistique de l'entropie, in: Max Planck-Festschrift, Berlin 1921, pp. 203-211.

Léon Rosenfeld, Introduction [to reprint of Bohr's Papers] in: Niels Bohr, On the Constitution of Atoms and Molecules, Copenhagen and New York 1963, pp. IX-LIV.

Roger H. Stuewer, William H. Bragg's Corpuscular Theory of X-Rays and γ-Rays, in: *British Journal for the History of Science*, vol. V, 1971, pp. 258–281.

Roger H. Stuewer, Non-Einsteinian Interpretations of the Photoelectric Effect, in: *Minnesota Studies in the Philosophy of Science*, vol. 5, 1970, pp. 246–263.

Bartelt Leendert van der Waerden, Sources of Quantum Mechanics, Amsterdam 1967.

Edmund T. Whittaker, A History of the Theories of Aether and Electricity, vol. 2, Edinburgh 1953.

Copies of the letters, manuscripts, and interviews constituting the "Sources for History of Quantum Physics" are deposited in the library of the American Philosophical Society, Philadelphia, in the library of the University of California, Berkeley, and in the Niels Bohr-Institute, Copenhagen. An inventory and report has been published in 1967: Thomas S. Kuhn, John L. Heilbron, Paul Forman, and Lini Allen, Sources for History of Quantum Physics, in *Memoirs of the American Philosophical Society*, vol. 68, Philadelphia 1967.

Name Index

Abraham, Max, 88

Bjerrum, Niels, 80n, 134–135, 141, 153
Blackman, M., 133
Bohr, Harald, 151–152
Bohr, Niels, 3, 45, 78n, 79, 81, 88, 94, 98, 99, 101, 128, 135, 142, 143,
Boltzmann, Ludwig, 2, 16–18, 22, 23–24, 64
Born, Hedwig, 134n
Born, Max, 65, 68, 87, 121, 125–126, 131–133
Bragg, William Henry, 127, 142
Bragg, William Lawrence, 127
Brillouin, Léon, 121
Brillouin, Marcel, 121, 140, 142
Broglie, Louis de, 143
Broglie, Maurice de, 139n
Bucherer, Alfred Heinrich, 66

Chwolson, Orest Danilovich, 30
Clausius, Rudolf, 5
Curie, Marie, 140, 153n

Debierne, André, 153
Debye, Peter, 42, 65, 75, 103–104, 107–108, 109, 110–111, 126, 127, 131–133, 141
Des Coudres, Theodor, 97
Drude, Paul, 34, 36, 150
Duane, William, 74, 127

Ehrenfest, Paul, 18, 20, 22, 60, 140
Einstein, Albert, 7, 16n, 20, 22, 26n, 31n, 34, 41, 42, 45, 46–47, 50–69, 73, 75, 79–80, 88, 89n, 90, 96, 106, 108–110, 120, 121, 124–127, 130–134, 135–139, 140, 147
Epstein, Paul S., 68, 80–81, 109–110
Eucken, Arnold, 140
Ewald, Peter Paul, 87
Exner, Franz, 91

Föppl, Ludwig, 157
Franck, James, 77–79, 127, 141
Friedrich, Walther, 127

Gans, Richard, 141
Gehrcke, Ernst, 77
Geiger, Hans, 22
Gerlach, Walther, 126–127
Gibbs, Josiah Willard, 5
Goldschmidt, Robert B., 137, 139n

Haas, Arthur Erich, 45, 61, 90–101, 109, 122, 141, 148, 149, 150
Hagen, Ernst, 35–36
Hallwachs, Wilhelm, 124
Hansen, Hans Marius, 157
Hasenöhrl, Friedrich, 34, 96, 100–101, 141, 153
Heilbron, John L., 68
Heisenberg, Werner, 24–25, 61n
Hertz, Gustav, 77–79, 127
Hertz, Heinrich, 124
Hevesey, Georg von, 146–148, 150–151, 152, 158
Hilbert, David, 69
Hopf, Ludwig, 141
Hund, Friedrich, 88n
Hunt, Franklin L., 74, 127

Jahnke, 12, 30
Jeans, James, 19n, 23n, 31–34, 37, 38, 42, 46, 47, 51, 54, 60, 91, 109, 111, 133, 139–140, 142
Joffé, Abram Fedorovitch, 124

Kamerlingh-Onnes, Heike, 141
Kármán, Theodor von, 131–133
Kaufmann, Walter, 66
Kayser, Heinrich, 6, 29–30
Kelvin, Lord (William Thomson), 65
Kirchhoff, Gustav, 5
Klein, Felix, 39n, 103
Klein, Martin J., 18
Knipping, Paul, 127
Knudsen, Martin, 140
Kuhn, Thomas S., 108, 155
Kurlbaum, Ferdinand, 12, 33

Ladenburg, Erich, 124, 125
Ladenburg, Rudolf, 67
Landé, Alfred, 121, 141
Lang, Victor von, 91

Langevin, Paul, 121, 141, 142
Langley, Samuel Pierpont, 6
Larmor, Joseph, 56n, 137, 140, 148, 151n
Laub, Johann Jacob, 51, 57, 61, 67, 110, 130
Laue, Max von, 74, 79, 105, 127, 128
Lecher, Ernst, 96
Leibniz, Gottfried Wilhelm, 1
Lenard, Philipp, 87, 124, 127
Lindemann, Frederick A. (Lord Cherwell), 130, 131, 134, 139n, 141
Lommel, Eugen, 30
Lorentz, Hendrik Antoon, 20, 31, 51, 54, 56, 58, 72, 75, 84, 87, 88–89, 91, 97–98, 109, 122, 137, 138–139, 140
Loschmidt, Joseph, 21
Lummer, Otto, 12, 15, 29, 30, 38, 39–40, 89

Mach, Ernst, 16, 50
Magnus, Alfred, 130
Markau, Karl, 125
Meitner, Lise, 67
Meyer, Edgar, 126–127, 141
Millikan, Robert Andrews, 125
Minkowski, Hermann, 39n, 67
Müllner, Laurenz, 96

Nagoaka, Hantaro, 87
Natanson, Ladislas, 18, 20, 141
Nernst, Walther, 45, 65, 108, 127–142
Nicholson, William, 81, 101, 143, 149, 150n, 151–152, 154

Oseen, Wilhelm C., 98
Ostwald, Wilhelm, 16, 50

Paschen, Friedrich, 6, 12, 29–30, 74, 88–89, 112, 141
Perrin, Jean, 50, 140
Planck, Erwin, 24–25
Planck, Max, 1–26, 30–34, 35, 37–39, 41–47, 53–54, 56–57, 59, 60, 62, 63, 66–68, 88n, 89, 91–92, 94, 98, 99, 100, 103, 105,

106–107, 108–109, 110, 113–118, 126, 134, 135–139, 140–141, 147
Pohl, Robert, 124, 127
Poincaré, Henri, 39n, 89, 121, 140–142
Prandtl, Ludwig, 133
Pringsheim, Ernst, 12, 15, 29, 38, 39–40
Pringsheim, Peter, 124, 127

Rayleigh, Lord, 8, 31–33, 42, 46, 47, 51, 54, 59, 91, 137, 138, 140, 142
Regener, Erich, 22
Reiche, Fritz, 26, 68
Richarz, Franz, 21
Riecke, Eduard, 30–31, 34, 157
Ritz, Walther, 87, 88
Röntgen, Wilhelm Conrad, 105
Rosenfeld, Léon, 98, 155, 156–157
Rubens, Heinrich, 8, 12–13, 15, 32, 33, 35–36, 130
Rutherford, Ernest, 21–22, 40n, 81, 87, 88n, 89, 128, 140, 142, 143, 146, 149, 151, 152, 154

Schidlof, Arthur, 99–101, 141
Schuster, Arthur, 137
Seeliger, Hugo von, 137
Seeliger, Rudolf, 77
Solvay, Ernest, 137–139
Sommerfeld, Arnold, 39–40, 42, 43–44, 45, 61, 65, 67, 69, 72, 75, 82–84, 87, 88–89, 97–98, 99, 126, 127, 129, 130, 131–132, 133, 141–142, 143, 149
Stark, Johannes, 41, 45, 56n, 60, 63, 68, 96, 104–107, 109, 127, 128, 140, 155, 156
Steubing, Walter, 74, 81, 104
Sudhoff, Karl, 97

Thiesen, Max Ferdinand, 12, 29
Thomson, Joseph John, 21, 56n, 87, 89, 92, 109, 128, 142, 148
Thomson, William, see Kelvin, Lord

Uhlenbeck, George E., 111

van't Hoff, Jacobus Henricus, 129
Voigt, Waldemar, 89

Waals, Johannes Diderik van der,
  137
Warburg, Emil, 80, 127, 141
Weiss, Pierre, 141
Whiddington, Richard, 149
Wien, Willy, 6ff., 12, 36, 39–41, 45,
  56, 59, 61, 74, 89–90, 91, 101,
  106, 128, 137, 140, 141, 152–153
Wiener, Otto, 96
Winkelmann, Adolph, 31
Wood, Robert Williams, 23–24, 26

Zermelo, Ernst, 2, 16, 22